Booker T. Whatley's Handbook On

How To Make
$100,000
Farming
25 Acres

Booker T. Whatley's Handbook On

How To Make $100,000 Farming 25 Acres

With special plans for prospering on 10 to 200 acres.

By Booker T. Whatley
And The Editors Of *The New Farm*

Edited by George DeVault
And The Editors Of *The New Farm*
Mike Brusko
Craig Cramer
Bob Hofstetter

Regenerative Agriculture Association, Emmaus, Pennsylvania

Dedication

This book is dedicated first to my loving wife, Lottie Mae Cillie Whatley, who has been my source of inspiration, pride and joy over the years. She was my "play mama" in college and, to this day, thinks I'm her little boy. I often wonder if I could ever go on without her.

I would also like to dedicate this work to my good friend Thomas S. Monaghan, founder and president of Domino's Pizza, who shares my belief in the goodness of small farms, and named his farm in my honor.

Booker Tillman Whatley

Acknowledgements

"How To Make $100,000 Farming 25 Acres" was the work of many, many people from Montgomery, Ala., to St. Louis, Mo., and Menomonee Falls, Wis., to Emmaus, Pa. Chief on that list is the staff and contributors, past and present, of *The New Farm* magazine: George DeVault, Mike Brusko, Craig Cramer, Bob Hofstetter, Ken McNamara, Frank Pollock, Teri Sorg-McManamon, Janice Chartrand, Gene Logsdon, Fred Zahradnik, Lesa J. Ayers, Dan Looker, Judy Yaeger and Pat Slattery. Special thanks also need to go to Michael Sands, Ph.D., director of agricultural sciences for Rodale International; Robert Rodale, chairman of Rodale Press and the Rodale Institute; John Haberern, president of the Rodale Institute; Thomas S. Monaghan, founder and president of Domino's Pizza; artists Charles Beck, Ed Courrier and Pamela Romano; Rose Reichl and Ramona Neidig of the Rodale Press Photo Library; Janet Glassman and the staff of the Rodale Press Library; Barbara A. Herman, Brenda J. Kline, Tom Woll, Pat Corpora and Ellen Greene of the Rodale Press Book Division; Skip Kauffman and Tom Morris of the Rodale Research Center; Al Johnson, farm manager of the Stony Brook Millstone Water Shed Association in Pennington, N.J.; Karl F. Ohm III, editor/publisher of *Rural Enterprise* magazine, and his Managing Editor Pete Millard; Carolyn Reich and her talented staff at the Type House in St. Louis, Mo.; Ward Sinclair and Cass Peterson of *The Washington Post* and Flickerville Mountain Farm & Groundhog Ranch in Dott, Pa.; Booker Whatley's many friends and colleagues; and, perhaps most importantly to Judith LeVan, Renae Rzonca, Lynn Kramer and Sue Snyder for keyboarding and processing the bulk of the manuscript.

Made possible, in part, by contributions of members of the Regenerative Agriculture Association of the non-profit Rodale Institute.

Design and Art Director: **Jan Georgen**
Cover Design and Art Direction: **Charles Beck**
Senior Production Coordinator: **Barbara A. Herman**
Typesetting Composition: **The Type House**

Printed in the United States of America on recycled paper, containing a high percentage of deinked fiber.

Much of the material in this book first appeared in slightly different form in *The New Farm*, magazine of regenerative agriculture; and Whatley Farms' *Small Farm Technical Newsletter*.

Regenerative Agriculture Association/ Rodale Institute

Chairman: **Robert Rodale**
Vice Chairman: **William E. Marshall**
President: **John Haberern**
Executive Director, Vice President: **James O. Morgan**

Copyright © 1987 by the Regenerative Agriculture Association of the non-profit Rodale Institute. All rights reserved. No part of this publication may be reproduced or transmitted in any form or by any means, electronic or mechanical, including photocopy, recording, or any information storage and retrieval system, without written permission of the publisher.

Published in the United States of America by the Regenerative Agriculture Association, 222 Main St., Emmaus, Pa. 18098.

Library of Congress Cataloging-in-Publication Data

Whatley, Booker T.
  Booker T. Whatley's handbook on how to make $100,000 farming 25 acres.

  Includes index.
  1. Agriculture—United States—Handbooks, manuals, etc.
  2. Farms, Small—United States—Management—Handbooks, manuals, etc. I. New farm. II. Title. III. Title: Handbook on how to make $100,000 farming 25 acres.
  S501.2.W47   1987   630'.68   87-20809
  ISBN 0-913107-09-3   hardcover
  ISBN 0-913107-07-7   paperback

  2 4 6 8 10 9 7 5 3 1   hardcover
  2 4 6 8 10 9 7 5 3   paperback

Table of Contents

Acknowledgements iv
Introduction vii
The Guru's 10 Commandments ... x

Part I The Theory

1. Take Charge Of Your Farm 1
   'City-To-Farm Roads' 1
   How To Avoid 'Gambler's Ruin' 8
      David Ehrenfeld
   Your Farm Is Worth More Than Ever ... 9
      Robert Rodale

Part II The Practice

2. Getting Started, Or Look Before You Leap 13
   Farming — A Matter of Dollars And Sense 13
      Judy Yaeger
   Know Your Cash Flow In Advance 14
   Let Your Customers Do The Harvesting 16
   Handling Money Safely 19
   Wamp's PYO Operation 19

3. How To Build A GUARANTEED MARKET, Establishing Your Clientele Membership Club (CMC) 22
   A Sample Letter For Prospective Customers ... 25
   Monitoring The CMC 28
   Annual Media Days 28
   Incorporating The Farm 29
   Teach A Cook To Cook 29
   Insurance — A Liability Policy Is Just Not Enough 30
   Say It with Signs 31

4. Equipping Your Whatley Farm, Sensibly And Economically 32
   3-Tractor Vegetable Growing System 34
   Bid Better, Buy Cheaper 44
   How To Buy A *Good* Used Tractor 47
   'Weatherproof' Your Farm With Irrigation 49
   Don't Guess, Soil Test 52

5. High-Value Crops And Enterprises 54
   Being Profitable Means Being Different 54
   These Little Lambs Mean Big Profits 57
   Your Own Hunting Preserve 60
   7 Times Better Than Beef 63
   Big Buck$ From Bizarre Breeds 64
   Freezer Beef And Lamb From A Few Acres 68
   The Kiwis Are Coming 69
   How To Grow Kiwifruit 72
   Money Does Grow On These Trees 74
   And Now, The Profitable Pecan 78
   Why Basil And Rosemary Ride The Subway 79

Nature's Hydroponic Harvest 81
  David McCoy
Growing Watercress Commercially 83
Triple Your Tree Profits 83
Finding Those Hidden Dollars In Your Woodlot 86
  Gene Logsdon
PYO Firewood 90
  David McCoy

6. Add Value To Everything 92
20 Times More Income From Wool 92
They Spell Profit, C-I-D-E-R 94
Double Early Produce Prices 97
Cash In On Raw Goat Milk Cheese 99
'Little House In the Woods' Is Retreat For
Vacationers 101
These Soybeans Earn ¥25,600/Bu. 104

7. Marketing, The Most Important Part 106
From Pasture To Plate, Direct Marketing Of
Livestock 106
He Bypasses 20 Middlemen 108
Target New Markets 112
'Billboard' In the Sky Boosts Sales 250% . . . 116
Manna From The Mail 118
Be A 'Price Maker,' Not A 'Price Taker' 121
Let Your Software Do The Selling 124
They Need Chemical-Free Food 126
Chemical Sensitivity Her Personal Nightmare . . 128

8. Beat The Weeds 130
Your Mower—The Best Herbicide Yet 130
Nature's Herbicides 132
Kick Up Your Yields 134

9. Beat The Bugs And Other Pests 137
A 'Natural' Answer To Antibiotics 137
Building A Better Fly Trap 139
A Guide To Fighting 40 Common Insect Pests 141
Other Fly Traps 142
Bug-Killing Cover Crops 142
Sticky Snakes! 144
Put The Bite On Snapping Turtles 144

Part III.   The Man

10. Conversations With Booker Whatley 147
We're All Workers In the Same Vineyard 148
The Real Test Of Human Behavior 150
'That's How Mama Always Did It' 152

11. 'The Ultimate Small Farm' 153

Appendices
I. Fencing Supplies, Information
   For Rotational Grazing 162
II. General Seed Houses 163
III. Vegetable Specialties 165
IV. Herb Seeds and Plants 169
V. Berry and Small Fruit
   Specialties 171
VI. Fruit and Nut Tree Nurseries   173
VII. People Who Can Help 176

Index 178

Introduction

You know the type: Their first reaction to any new idea is always negative.

"It won't work!"

"What do you want to do a dumb thing like that for?"

"It'll never fly!"

Booker T. Whatley generates a lot of comments like that with his plan for making $100,000 by farming 25 acres.

'Nothing Ventured...'

Although hundreds, maybe even thousands, of farmers throughout the United States and Canada are now using parts of Whatley's plan to earn more money from limited acreages, the critics still aren't satisfied.

"Well!" they snort indignantly. "Of course you know he hasn't tried it anywhere outside of Tuskegee Institute... and he had a Rockefeller grant for that work. We don't know how it will fly in the real world."

While working with Whatley on this book, I've heard those and other derisive comments many times during the last year. At first, cracks like that bothered me. I soon quit worrying about the critics, though. It quickly became obvious from the tediousness of their complaints that they had blessed little in the way of imagination, vision or inventiveness.

This book is not written for the critics. It is meant for farmers and other practical people with the ability to quickly analyze, adapt and implement new ideas.

Common sense dictates that there are few, if any, ironclad guarantees in farming, other businesses or even life, for that matter. More often than not, what you get out of something depends on what you put into it. Success depends almost entirely on the intelligence, drive and resourcefulness of the individual, and the ability to recognize and take advantage of opportunities as they arise.

The critics conveniently seem to forget all of that, plus the fact that our present farming system isn't working terribly well. In fact, it's failing miserably for many people. We are desperately in need of new answers. So what if they're "scientifically" unproven or run counter to conventional wisdom? They just might work.

'Reap What You Sow'

"Sure as hell couldn't do any worse than a lot of farmers are doing today!" Whatley growls.

The printed word simply does not do his voice justice. Whatley's whiskey baritone rises and falls, twists and turns like the pitch of a rolling country road. Even when he's barely whispering, Whatley speaks with the passion of a Baptist preacher flaying his flock with threats of hellfire and damnation. His Southern accent, as dark and thick as molasses, often plays tricks on Yankee ears. The sweet smoothness of that accent is made raspy, though, by the roughly two packs of unfiltered Lucky Strikes he smokes every day.

The needle on Whatley's speedometer pushes 65 as his black Grand Am tops a hill in the curving two-lane road leading from Tuskegee back to Interstate 85 and Montgomery. All the windows in the car are down. Whatley is holding a book of matches and the wheel in his left hand. Despite the rushing wind, he strikes match after match in no particular hurry to get them to the unlit cigarette dangling from his lips, as he drives and talks and gestures.

"I was talking with a farmer up in Michigan and he said, 'It took me 12 years to learn to grow soybeans. And here, you want me to grow 10 different crops? Why . . .' The words come faster and faster until they're just a blur. His voice soars two whole octaves, then breaks into a delighted cackle. 'That'll take me 120 years!'

"That farmer just didn't want to change!" Whatley snorts. "Why, you can learn everything you need to know about blueberries in 15 minutes. Hell, make that five minutes. Blueberries need a pH of 4.5 to 5.5. You can't grow 'em without it. They need irrigation. If you're too poor to irrigate, you're too poor to be messin' with blueberries. Space them 6 feet apart in the row and the rows 12 feet apart, so that you have 605 plants per acre. You need to plant different varieties so that you have production through the whole season."

Bang! Just the facts. That's Whatley in a nutshell, his approach to farming and the basics of life. Sure, Whatley likes to joke around as much as anyone. He has a delightful, somewhat impish, sense of humor. But when it comes to farming, he's all business. He has no patience or sympathy for people who won't do what needs to be done.

Do It Right, Or Not At All

Part of that comes from his upbringing. Whatley is the oldest of 12 children. He was born in Anniston, Ala., on Nov. 5, 1915. That was the year that Booker Taliaferro Washington died. Although his middle name is Tillman, Whatley was named after Washington, the founder of Tuskegee Institute. Whatley has a special place in his heart for Washington and Tuskegee. He especially likes the famous statue of Washington at the heart of the Tuskegee campus. It shows a resolute Washington ceremoniously drawing a dark cloak from another black man's eyes.

The inscription at the base of the statue reads:

He lifted the veil of ignorance from his people and pointed the way to progress through education and industry.

"My granddaddy and great grandparents were born in slavery, never attended any school, but each had a head full of common sense," Whatley says of his family. "They had tremendous influence on my values, opinions, drive and self-confidence. From them, I learned that everybody respects a man with backbone and that you should stand up for what you believe in."

That, and his distaste of pizza, is why Whatley is apt to order a steak or lobster rather than pizza when he dines with his friend and small-farm consulting client, Thomas S. Monaghan, founder and president of Domino's Pizza (see "The Ultimate Small Farm" on page 153). It is also why Whatley now receives $2,000, plus expenses, for speaking engagements. He started out charging just $500 a few years ago. Speaking requests kept coming, so he doubled his fee. Requests still kept coming, so he doubled his fee, again. He now makes about 20 personal appearances a year.

Whatley's "take charge" attitude also comes from his more than 30 years in the military. He flew fighter planes over Italy during World War II. In early 1952, the Army recalled him for active duty in Korea. Maj. Whatley was placed in charge of integrated troops for the first time. The admonition from his white commanding officer was, "Go up there and be as hard as you damn please, but be fair." That advice stuck with him. His other motto is "Find the good and praise it."

Whatley learned to know—and respect—the laws of nature in college, from his undergraduate work at Alabama A&M through his doctorate in horticulture from Rutgers University. "If you want to grow blueberries," he says, "here's what you have to do . . . If you can't or won't observe the basic laws of horticulture, then forget it! Either do it right, or don't do it at all, because the results just won't be what you wanted. You'll simply be wasting your time and money."

Despite his strict adherence to fundamentals, though, Whatley is not as inflexible as he might seem. For example, when I first made Whatley's acquaintance in late 1981, he laughed at the idea of farming without chemicals. Like so many professionals brought up in the land grant university system, he believed only in following university-recommended spray schedules and freely recommended them. But as growing numbers of consumers became concerned about pesticide residues in their food, Whatley started advising farmers to reduce or even eliminate the use of many chemicals. Today, he says that is one of the most compelling marketing tools a farmer can use.

It's Up To You

Like Whatley's small-farm concept and plan, his book is not for everyone. What you get out of it is strictly up to you. When Whatley first proposed this book, we decided that it should not try to be an encyclopedia of specific production practices. Enough books have already been written on how to grow blueberries, sheep and just about every component of a Whatley-style farm. Although filled with practical how-to information, this is mainly an idea book. It is meant to make you question the established way of doing things and start thinking on your own.

If you're serious about this type of farming, you have a big job ahead of you. And your work is just beginning. You must take the initiative and become a true student of high-value, pick-your-own crops and livestock enterprises. If you're counting on just this one book to give you all the answers, then forget it.

It is the rare individual, indeed, who can do everything Whatley recommends. For starters, you have to have the right personality. Does Barbra Streisand's song "People" describe your overall feelings about your fellow human beings? If not, find another line of work. There is no way someone who is always an old grump is going to be able to successfully recruit and keep up to 1,000 member-families in a Clientele Membership Club.

Whatley is the first person to admit all that. "Nobody that I know of is following my concept 100 percent. I don't think there ever will be. I hope not. Americans are independent and farmers are even more independent than most of us. If everybody started doing just what I say, it would mean that they aren't using their heads enough and coming up with enough ideas of their own. They'd be relying too much on someone else's formula or recipe. That's just what got farmers into a lot of the trouble they're in today."

Farmer George McConnell agrees. "It's quite a job," he says of trying to implement the Whatley plan 100 percent. "Most people are not qualified to carry it out." McConnell, whose son now operates the family's 220-acre U-pick fruit and vegetable farm northwest of Mt. Vernon, Ohio, used to be a regular contributor to Whatley's *Small Farm Technical Newsletter*. But after two years, he quit writing about how to grow blueberries, strawberries and other crops.

"I felt I was getting some people into trouble," he says. "If a person doesn't have the specific things Dr. Whatley recommends from the start, he's in trouble.

"When it comes to the membership club, most people are not capable of handling that kind of business. You need better-than-average salesmanship. If you have it, maybe you should think about selling something else."

McConnell found that he couldn't implement a Clientele Membership Club on his farm, which had been open to the public since 1968. "It's tough to charge your existing customers a membership fee or tell them that they can't pick at your place anymore," he says. "Start your club as you start your farm."

On the agronomic side, McConnell says, "the key is top management. They either have it or they don't." This summer, McConnell took out 11 of his 13 acres of grapes. He says they never really paid for themselves. McConnell advises long and careful study before getting into any new crop in a big way, especially those with high start-up costs.

'Lifting The Veil'

By now, if I've done my job properly, you have a more complete understanding of Booker T. Whatley—both his small-farm plan and the man, himself. There is a lot more to Whatley and his ideas than first meets the eye with a headline about making $100,000 from 25 acres. Read on. You'll find many more helpful insights and a wealth of practical—and profitable—ideas in this book.

Meantime, whatever you think of Whatley and his ideas, I think you'll have to agree on one thing: He is making us all reexamine our present farming system, and think much harder about what went right, what is not working so well and what we can do to build a better future. In the process, Whatley is lifting the "veil of ignorance" a little higher from each and every one of us.

George DeVault, Editor
The New Farm
Sept. 3, 1987
Emmaus, Pa.

'The Guru's' 10 Commandments

thy small farm shalt:

I. Provide year-round, daily cash flow.

II. Be a pick-your-own operation.

III. Have a guaranteed market with a clientele membership club.

IV. Provide year-round, full-time employment.

V. Be located on a hard-surfaced road within a radius of 40 miles of a population center of at least 50,000, with well-drained soil and an excellent source of water.

VI. produce only what thy clients demand—and nothing else!

VII. shun middlemen and middlewomen like the plague, for they are a curse unto thee.

VIII. consist of compatible, complementary crop components that earn a minimum of $3,000 per acre annually.

IX. be 'weatherproof,' at least as far as is possible with both drip and sprinkler irrigation.

X. be covered by a minimum of $250,000 worth ($1 million is better) of liability insurance.

Part 1
The Theory

Chapter 1
Take Charge Of Your Farm

'City-To-Farm Roads'

Adapted from a speech by Booker T. Whatley to the Conference on Sustaining Agriculture Near Cities sponsored by the Tufts University School of Nutrition, Boston, Mass., Nov. 20, 1986.

Ladies and gentlemen, I want to say at the outset that the group of farmers that I am trying to assist is the same group of farmers that USDA, the land grant college system and the Cooperative Extension Service told 25 or 30 years ago to, "Get bigger or get out! You go to town. You get a job." Some of these guys went to town and got a job, like they were told. *And they kept these farms.* Right now, the small farmers in this country are in the least financial trouble of any group of farmers in the United States.

Now if you listen to the press, you think it's the other way around. But what the press and the media is always talking about is the "family farm." You don't hear a thing about small farms. You only hear about "family farms." I'm talking about small farms, the group that the USDA kicked out 30 years ago.

Let me just say this: Just in case anybody thinks I'm anti-big farm, I am not anti-big farm. I was educated in the land grant college system. I've worked in it more than 30 years. All I'm advocating is saving 100,000 of the farmers that USDA, the land grant college system and the Cooperative Extension Service kicked out 30 years ago. I get along well with the Farm Bureau, the Grange, the Farmer's Union. I'm not anti-big farm. Maybe I come across like that, but I'm not. I recognize three categories of farms in the United States. I recognize small farms as farms that consist of from 10 acres to 200 acres, and family farms of from 200 to 1,000 acres. Anything above 1,000 acres is a large farm. It is also my position that this country can support all three categories and neither should be hurt by the other.

"Get bigger—or increase production on the same amount of land. Or get a job in town," is the advice still being doled out by the likes of former Agriculture Secretary Earl Butz.

My quarrel is with the land grant college bunch, because they very seldom have a really new idea, let alone one that's going to do something good for the farmer. They all think exactly alike. Why is that? They were trained to think that way. They all went to the same schools and listened to the same bunch of professors who had done exactly the same thing when they went to school. It's like a big social club, a fraternity. And everybody thinks everybody else is just great, because they all think the same way.

How do USDA and the land grant colleges view the small farmer? If he's white, he's perceived to be wearing an old hat, faded overalls with a few patches, and chewing tobacco, with the tobacco juice running out of one side of his mouth. If he's black, he's perceived to be wearing an old, beat-up hat, faded overalls with patches on the patches, and he must split every verb that comes out of his mouth. When they say "limited-resource" farmer, they're not just talking about a farmer who has limited land or limited financial resources. They're talking about a guy that's limited in the head!

Now my peers and opponents say the Whatley Small Farm Plan is just too complicated for small farmers in this country. My most vocal adversary is a professor of economics at Oklahoma State, Luther Tweeten. He says that only a handful of small farms are going to be saved by my concept and plan. Now I only propose saving 100,000 small farms by 1995. But Luther Tweeten says we're only going to save a handful.

My reply to the people like Luther Tweeten has always been that I agree with you. My plan *is* complicated. It requires good management and all. But I also add that my granddaddy, who was born in slavery and never went to school, operated a farm 10 times more complicated than this. But one thing the old man had going for him was a head full of common sense. Now, you see, we Americans live under all kinds of myths. And one of them is that we correlate—positively correlate—the number of degrees an individual has with the amount of common sense he has. I would like to inform you that there is absolutely no correlation, whatsoever. None!

Another of the critics says, "Yes, but . . . your plan doesn't help *everybody*." Well, I'm not trying to help everybody. I was in Des Moines earlier this year and there was a guy there from Cornell. He said, "What are you going to do about the people in Harlem?" I said, "Nothing! Hell, I'm no sociologist, man. I'm a horticulturist." Then some other joker says, "Well, what if *everybody* gets into this?" Everybody can't get into this! You've got to be a good manager to operate one of these farms. You've got to be a good planner and you've got to think for yourself. See, too many farmers in this country spend too much time on the seat of a tractor. And some of them even have those things air-conditioned. They've got televisions in them and tape players. What they really need is an air-conditioned office where they can do their planning and thinking and managing. Tractor drivers are a dime a dozen.

Farmers in this country are hard-working. But somebody has to tell them that they have to be hard planners, hard managers, and hard thinkers, too. There's no reward for just hard work, really. There's nothing.

I would also like to inform farmers that it is soon going to be much more profitable to farm their land and manage the commodities they produce than to farm the government. Now at this point in time, it is just more profitable to farm the U.S. government than to farm your land. And all of the successful farmers in this country at the present time are not farming the land, they're farming the government. But that can't—and won't—last for many more years.

Now you know that the Democrats gained control of the United States Senate on Nov. 4, 1986. This does not mean that the farm problems of this country will be solved overnight. I'm a Democrat, but I'd just like to add that the probability is good that they're *not* going to be solved. It is my position that the farm problems of this country will be solved when, and only when, mismanagement is made a federal crime. They will also be solved when we pass the 27th Amendment to the United States Constitution. And that 27th Amendment has got to be very similar to the First Amendment to the Constitution, which has to do with the separation of the state and church. Now some kind of way, we are going to have to get our government out of agriculture. The way to do it is pass the 27th Amendment—and really separate them.

Most Americans worry about big, cooperative-type farming taking over food production in this country. I only worry about the day that the United States government takes over food production in this country. I think then, and only then, we will be in great danger and face possible starvation for all of us. Too many agriculturists or too many agricultural experts labor under the myth that the way to solve the problems facing farmers in this country is to set up state-owned and -operated farmers markets, cooperatives and roadside stands. The only thing that's wrong with those outlets is that each requires small farmers to do something that they cannot afford to do: And that is to harvest

So-called farmers markets don't help small farmers one bit, because they force farmers to do so many things they just can't afford to do.

produce and go through that harvesting routine of picking, washing, grading, packing containers and then transporting the produce to market 200 miles away. He can't afford to do it. So nothing could be further from the truth than to think that cooperatives, farmers markets or even roadside stands are going to solve the problems. *Small farmers just can't afford to do all that.*

And that's why I have a whole lot of trouble accepting the national cliche of what we call "Farm-to-Market Roads." What I'd like to see them become is "City-to-Farm Roads," roads that city folks would use to come out to area farms and harvest food for their families.

The day is fast approaching when there will be no successful small farmers in this nation peddling in parking lots, church yards, on streets and street corners and in state-operated regional and local farmers markets, exhibiting their wares in 100 degrees and/or freezing temperatures.

What I'm going to talk about is the small farm plan being recommended for staying in agriculture near cities. This plan involves what I call "*The Guru's 10 Commandments.*" But first, let me tell you how I came up with that title.

Ward Sinclair of *The Washington Post* wrote an article about my concept and plan a few years ago. And he said that, "A lone guru in Alabama has found a way for small farmers to make $100,000 from 25 acres." To tell you the truth, I guess I'd heard of the word guru before, but I had no conception whatsoever of what a guru actually was. At that time, my wife, Lottie, was the nutritionist for the Montgomery Head Start program. There was this Indian lady in the office, and Lottie carried Ward's article down there to show her. And this Indian lady said, "Ooohh! Is your husband a gu-ru?" And she raised so much sam that when Lottie got home, I said, "We better look that word up and see what that writer called me."

So, anyway, as a result of that, I came up with what I call "The Guru's 10 Commandments." If you're going to assist small farmers, you'd better get used to abiding by these principles. And let me just say that these 10 principles apply all over this country. Only the crop mix will vary. Now we try to make a whole lot about this crop-mix thing, but there's not that much difference there. For example, everybody in this country eats strawberries. The few who don't are allergic to them. You don't have to worry about those people. The same holds true for blackberries and raspberries, blue-

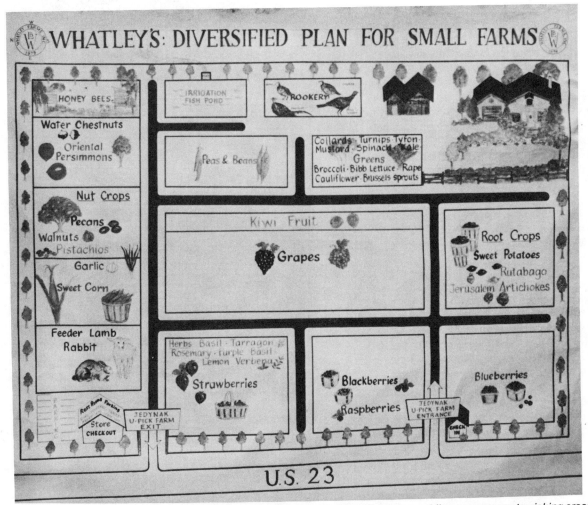

The well-designed Pick-Your-Own farm has one entrance and one exit, with roads providing easy access to picking areas.

berries and root crops. Now I give seminars all over this country. If I was talking to somebody in the South, I'd be talking about sweet potatoes. But when I'm talking to the Yankees, I have to talk about rutabagas and Jerusalem artichoke and some of these other root crops.

Now all of us eat greens. But, here again, down South you're talking about collards and turnips. Up here, you're talking about spinach, kale, broccoli, brussels sprouts and this kind of stuff. But all of us who can afford to . . . eat quail and pheasant. All of us like fish. All of us like honey and bee products. You can grow Chinese chestnuts up here, nuts, sweet corn, and lamb. So there's not that much difference in the eating habits of the American people. And the biggest thing is with the root crops and with the grains. All of us eat peas and beans.

I just want to point out if you're a small farmer producing peas and beans, for God's sake, have yourself a sheller. Now the middle-class American housewife doesn't mind coming out with the kids and picking a few bushels of peas. But she doesn't want to spend the next six months shelling the things. So offer a shelling service.

OK, so let's run through "The Guru's 10 Commandments."

First, all *small farms must provide year-round, daily cash flow.* Year-round, daily cash flow is designed and built into this small farm. One thing that you'll notice is that this is a diversified plan. Diversification does a couple of things for you. Number one, it provides daily cash flow. And number two, it protects you against farm failure. Now I was talking last week with a gentleman from Michigan. At just the time his strawberries were getting ready, they had 7.5 inches of rain and wiped out his strawberries. I recommend that these types of farms have a *minimum* of 10 components, each representing no more than 10 percent of the gross of that farm. So if you completely lose one component, you're still operating at 90-percent efficiency. And there are not many things in this country that operate at 90-percent efficiency. These are all high-value crops and enterprises, each earning a *minimum* of $3,000 per acre *per year*. A lot of the crops and enterprises that we recommend will far exceed that minimum.

A diversified small farm with year-round, daily cash flow has a greatly reduced need to borrow money from FmHA and other lending agencies who are most reluctant about lending to small farmers in the first place. FmHA's relationship with small farmers is something like hiring a fox to guard your chicken coop. Maybe it's more like sending a rabbit to the supermarket to buy lettuce.

The next principle is that *each one of these farms must be a pick-your-own operation,* because that greatly reduces the cost of the labor component. A PYO operation requires excellent management and detailed planning. Any successful PYO operation has these features:

• *One place where you enter this farm and one place where you exit.*

Now the husband should greet these ladies when they come onto this farm, supervise picking and teach them how to harvest, and this kind of thing. But, for God's sake, put your wife in charge of where they exit and pay that money. Now I put a lot of emphasis on that, for this reason: If you men get over here at the exit, some of those charming women are going to come through here and jive you right out of your strawberries. So put your wife in charge of that checkout station.

• *Parking should be on both sides of the road near the commodity being harvested.* Don't build a central parking space on this farm and transport your clients to the crops to be harvested. Let them drive up to or near the crop that's going to be harvested.

• *Provide comfort stations,* both men's and women's, in the area of whatever you're harvesting. And some farms even have telephone jacks out there so that their clients can call home and leave word where they are and when they'll be home.

Hay wagons may be OK for field days, but there is too much labor and liability involved in using them to haul U-pick customers to and from your fields.

- *Do not allow pets of any kind on this farm.* No pets, whatsoever, because folks will bring dogs and some of these dogs have a way of getting in heat. Then you have half the dogs chasing the females and the other half of them fighting. So don't allow them on the farm. No dogs. No pets, whatsoever.

- *It's not necessary with these kinds of farms to advertise.* Advertisements cost money and publicity doesn't cost anything. Plus, when we get into the clientele membership aspect, you're going to see that there is absolutely no need for you to be buying radio time or television time.

Now the third principle is that *each small farm must have a Clientele Membership Club (CMC).* That's because with the way PYO farms are presently operated in this country, you have absolutely no control over who is coming. You don't know who is coming. You get on the radio and folks just pile on you, or they don't come at all. But with a CMC, it gives you the control and precision that you need and want in this kind of operation. A CMC consists of a group of loyal people, clients, patrons or customers. The clients perceive the American small farmer with a CMC as being a family man, religious, respected in his community and a dedicated individual who is engaged in the business of farming to provide them with wholesome, high-quality, *and contamination-free* fruits and vegetables, fish and game birds and their products, honey bees and their products, nuts, rabbits and their products, care and feed for their horses, and recreation for their families in a rural setting.

Middle-class American families are prepared to pay for these goods and services. The small farmers of this country are in an excellent position to totally exploit the middle-class American housewife's belief that the minute you turn any food over to a food processor, they are going to put

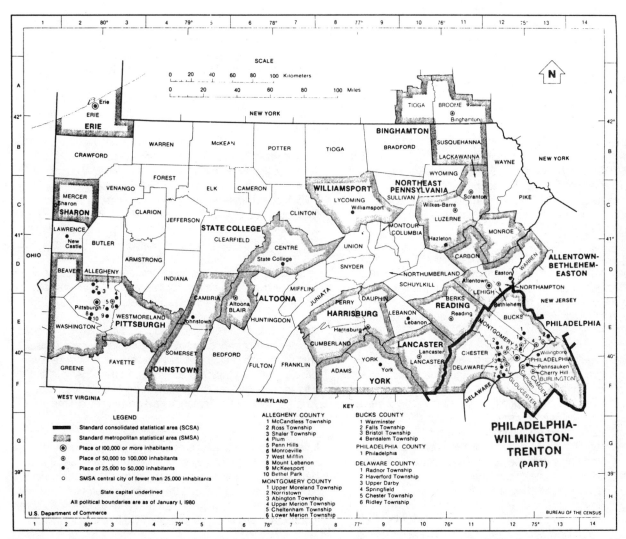

Shading indicates Standard Metropolitan Statistical Areas with populations of 50,000 or more. Your Whatley-style farm should be within a radius of no more than 40 miles of one or more of these centers. Maps such as this are available for every state from the Bureau of the Census of the U.S. Department of Commerce.

something in here that's going to kill you. They believe that. Whether you believe that or not, they believe that, and they have good reason for believing that. All you have to do is look at the recent EDB incident, the estrogen incident, and on and on.

You hear folks talking about convenience. It's a myth. These women don't want convenience. They want to be able to go on a farm, harvest *contamination-free* fruits and vegetables themselves, take it home, process it themselves and put it in their freezers. What they're looking for is good nutrition for their husbands and their kids. It is *contamination-free* and *high-quality* food products that these ladies are looking for.

So it becomes the duty and responsibility of the small farmer to seek out people, mostly city folk, to become members of his CMC. Now you have outfits in this country that will set up your CMC for a fee, and it's usually the $25 annual fee that we recommend that you charge. You should not allow that to happen to you. You should do it yourself. Because what we're talking about is a small farmer who knows his clients on a one-to-one basis. His clients, the housewives, respect him. The farmer is their buddy. These ladies feel like this farm is their farm. You should encourage that feeling. So don't allow anybody to set up your CMC for you.

The other reason you don't want anybody to set it up is that the CMC is not only your market, it's your GUARANTEED MARKET! One of these farms, a 25-acre farm, will support the freezer needs of 1,000 households. So that means the CMC should consist of 1,000 households. And I repeat, it's not only a market. It's your only market. IT'S YOUR GUARANTEED MARKET!

People say, "Oh, it would be so hard to get 1,000 of these people." My reply to that is, well, everything is hard! Life's hard! I've been married so long that I can't remember when I was single. It's hard to stay married, you know. I belong to the Baptist Church. It's hard to stay in the church, you know. Everything is hard.

But where you can provide yourself with a guaranteed market, it's worth it. It is worth it! So when the farmer gets on the phone, or however he's doing it, and starts talking to these ladies, the first thing to ask is, "Do you have a freezer?" Now if the lady says, "No, we haven't bought one yet," you can get off the phone with her, because you're wasting time. She doesn't have anywhere to put anything. And some of these housewives, believe it or not, not only have one, some of them have got two or three freezers. You're trying to find people to join your CMC who have the same philosophy as the ant. They believe in preparing for the winter. They believe in it, religiously. *The CMC is the lifeblood of any diversified small farm operation.*

Now to belong to a Clientele Membership Club, we recommend that the farmer charge an annual fee of $25 per year, payable on the first day or second day of January. What that does is give your clients the privilege to come on this farm and harvest produce at *60 percent* of what your local supermarket is selling the same item for. If the supermarket is selling its blueberries for $1 you sell 'em for 60 cents.

Sell everything by weight. Don't sell anything by volume. Sell it all by weight. The reason you want a PYO operation is because these city folks will pick just about anything. They'll pick produce that the farmer would never dream of picking. And you want to sell by weight, because they'll pick sticks and stones, too. I recommend to farmers that they don't pick anything out of what their customers pick. They want rocks in it? Leave 'em in and weigh it.

When you build a pond and stock it with fish, you can also sell fishing rights. Then the CMC fee goes to $40, but your members are still only paying 60 percent of what you're letting the other guy fish out of your pond for. Down our way, that ranges from $1 to $2 a pound for the first fish they catch.

Now, the fourth principle is that *the farm must provide year-round, full-time employment*. You just invite disaster to set up one of these farms and think you're going to work at it part-time, or you're going to be an S-S farmer, a Saturday and Sunday farmer. You've got to work this full-time.

The fifth principle involves where to locate this type of farm. Now normally, what I have when I give these talks is the easel up here with this data to show the Standard Statistical Metropolitan Areas (SMSA) in that state. But since we got so little response from New England, I didn't even have my girl make up one for New England. *This kind of farm needs to be located no more than 40 miles from a SMSA, or population of 50,000 people or more.* These city folks don't mind driving up to 40 miles. But after that, they are taking a trip, you know, and they're not interested in that. They're not interested in taking any trip to pick blueberries. But they will drive out to pick blueberries. *Your farm needs to be on a hard-surfaced road with a good source of water, because every square inch of this farm is under irrigation. You must have well-drained soil. That's where you locate.*

What should small farmers produce? *Small farmers should produce what their clientele demands and nothing else!* That's the sixth principle. Now farmers are notorious for producing what they *like* to grow and then trying to sell it to somebody. I mean they are good at that! So produce what your clientele demands and nothing else. I mean nothing else!

If your clientele demands collard greens, you shouldn't waste your time trying to educate them to eat kale. If they demand collard greens and turnips, that's what you grow. Nothing else. And I'm sure the school of nutrition here will bear me out that changing the eating habits of folks is some kind of task, anyway. You just don't do that a whole lot.

The next thing is small farmers must completely eliminate middlemen and middlewomen from their farming operation. And you accomplish that with this CMC. See, when some folks talk about direct marketing, they mean farmers doing all of the harvesting and hauling it to a farmers market. But when I use the words direct marketing, I mean a situation where the clients come to the farmer, do the harvesting and take it

You must eliminate all vestige of middlemen and middlewomen from your farming operation.

home with them in their own containers. The small farmer cuts down on his cost of production and gets 100 percent of that housewife's dollar spent on what he grows. One of the problems facing agriculture in this country today is that the poor farmer is getting such a small share of the housewife's food dollar. Under our system, the farmer sets his price! HE sets his price! The large farms in this country have never enjoyed the luxury of setting the price of anything. So the small farmer must eliminate the middlemen.

Next, *the farm must be protected against weather.* We do that with irrigation. This protects us against drought, late spring frosts and early fall frosts. So the farm has to be under irrigation and we call that weatherproofing the farm.

Finally, after you have done everything right on this farm, then you *cover yourself with a minimum—and I mean a minimum—of a quarter of a million dollars worth of liability insurance.* Now I talk about city folks and how dumb they are, and a lot of it is true. The reason you want a pick-your-own thing is these city folks will pick sticks and stones and stuff the farmer would never dare pick. But there is one thing that city folk know how to do. *They know how to sue you.* So you have to keep that in mind.

OK. Now, let me close. For the last 15 years, I have been championing the cause of small farmers in this country. For a long time, I was a lone voice in the wilderness, even though, since 1980, I have enjoyed excellent media coverage. But it was only this fall, in 1986, that people really started to listen. Now, it seems like they just can't hear enough about small farms. To get a better idea of that interest, just listen to my speaking schedule for this fall. On October 23-24, I was in Sumter, S.C., as a guest of the commissioner of agriculture. On November 12-13, I was in Kansas City as a guest of *Acres U.S.A.*'s annual farming conference. Today, November 20, I am in Boston as a guest of the School of Nutrition of Tufts University. And on December 2, I will be in Des Moines, Iowa, as a guest of *Successful Farming* magazine's ADAPT 100 (Agricultural Diversification Adds Profit Today) conference. They have already registered more than 5,100 small farmers from across this country. It will be the largest gathering of small farmers ever assembled in this country. So there are a lot of people interested in this sort of thing, whether the so-called experts and agricultural officials endorse it or not.

Professionally, I think that those of us who are genuinely interested in sustaining agriculture near cities need to constantly keep all of those changes in mind, and guard against doing what I sometimes accuse the land grant colleges and USDA of doing: taking up a prominent position of leadership—from behind.

Thank you very much.

Questions

Q: *I agree with you and 95 percent of what you're saying, but there are a couple of points that you could help straighten me out on. I'm doing something very similar to this. One of the things I really didn't like a whole lot was the idea that we ought to be charging 60 percent of what the supermarkets are charging, when we're offering a much better quality product. We're also offering entertainment, a day out in the country, recreation for the kids. My feeling is that we probably ought to be charging 60 percent more than the supermarkets.*

A: About 50 percent of the farmer's cost of production is tied up in harvesting, and these people are doing that for him.

Q: *But the other part that bothers me is that you're depending on a group of people that are interested in freezing and canning. Even national companies like General Foods, who makes Sure-Jell brand of fruit pectin for thickening jams and jellies, say that the numbers of people who are interested in canning, freezing and jamming are really dropping off very quickly.*

A: Yeah, but with my concept, YOU pick your clientele. You pick the bunch that wants to freeze. If you're going to pick people who want to do something else, you can't blame anybody but yourself. You're selecting these people.

Q: *If you don't advertise, how does the farmer who is just starting to diversify under your system get customers?*

A: Well, every community has a phone book. And, when you're dealing with a town of 50,000 people, there are a lot of office buildings that will be listed in the phone book. Now the small farmer can go to these buildings, talk to the building superintendents and ask to put his sign on the bulletin board on each floor. These signs should be neat and to the point. They should have the farmer's telephone number and all the details on the farm. People see these coming and going from work and during the day. Some of them call you, and we get word of mouth going.

Q: *Do most of your pickers come on Saturday and Sunday? I go every day. I go by and pick my supper on the way home. It's like going to the supermarket.*

A: That's right. That's what happens here. The farmer says he's going to be open from 9 a.m. to sundown or from 10 a.m. to sundown. He and his clients work that out. It's not just one day a week or something like that.

Q: *In the beginning, when you're first starting up, I'm not quite sure how you're going to get people to pay $25, because you're breaking them out of such a habit that they have of going to the U-pick when they want. Can you do this by having things priced at $1 a pound, 60 cents if you're a member? Can you have a two-scale pricing?*

A: We don't mess with people who are not members. We're not concerned about them. We want to know who's coming. We want control!

That's why you have to have a computer. Every time your clients come to your farm, you want to record when they come and how much of each thing they buy. You want to know the per-capita consumption of your client households on each item you sell. I don't think farmers in this country keep records, because, if they did, some of them wouldn't be doing some of the things they do. I know it. This requires good record keeping.

And once you explain to folks about how the Clientele Membership Club works and the fact that for their $25 they'll be getting fresh-picked and contamination-free produce for 40 percent off the supermarket price, there shouldn't be any problem getting them to pay the annual membership fee. Everybody likes to belong to something, especially Americans. We're real joiners. And everybody likes to save money. With the Clientele Membership Club, they can do all of that at once.

How To Avoid 'Gambler's Ruin'

DAVID EHRENFELD

In the 1970s, when the price of oil jumped, farm machinery and chemicals became more expensive because of the energy required in manufacturing and processing. For instance, the production of nitrogen fertilizer, the principal chemical input in farming, needs enormous amounts of energy. And accompanying the higher oil prices was an upsurge in interest rates.

To keep farming the way the system mandated — and the only way many farmers now know — it was necessary to borrow increasing sums of money. The inflexibility of the system was striking. For example, some California banks — closely associated with petrochemical companies — insisted on approving farmers' pesticide-spraying protocols before authorizing loans.

By the late 1970s, as farms began to fail, land prices in most of the nation's agricultural districts fell. Thus, the value of land as loan collateral declined. Many farmers, with up to 70 percent or 80 percent of their assets leveraged, found that interest on loans was a major expense. Farm profits were decreasing, despite higher yields and sales volume. Finally, the increased supply worldwide led to falling commodity prices.

The Gambler's Ruin

This situation could not continue. The reason is clear to anyone who knows probability theory. Its familiar "gambler's ruin" paradigm, described more than a century ago, states that in a fair game of chance with 50-50 odds, the house will eventually beat the gambler most of the time. This is because the house has far greater assets. In the normal ups and downs of the game, the gambler is much more likely to go broke first. Gambler's ruin usually comes quickly when the stakes are very high. If the bets are kept small, the game is likely to last much longer.

The farmers now facing bankruptcy are those who have annually risked between 25 percent and 40 percent — or more — of their assets to borrow the money to cover costs. But not all farmers in the United States rely on costly inputs. For many years, the Amish have practiced low-cost farming, often achieving yields comparable to or exceeding those of their neighbors. Recently, the Amish have been joined by an increasing number of organic farmers. For both groups, costs usually do not exceed 10 percent of assets.

In a simple model of the economics of conventional farming, a computer was programmed to bet one-third of a gambler's starting assets in each trial, at 50-50 odds. In 10 runs, the gambler went bankrupt every time. The longest run was 118 trials; the shortest was 6.

When I substituted a bet of 10 percent of assets for that of 33 percent in the computer model, the results were strikingly different. In 10 runs, the shortest was 48 trials. And 3 runs were stopped at 2,000 trials, with the gambler still in the game.

Another variable is the diversity of crops planted. Conventional farmers tend to grow only one or two crops, which makes them highly vulnerable to the vagaries of weather, epidemic disease, insect pests, and market fluctuations. In contrast, the most successful low-input farms tend to be highly diversified. Their profits cannot be destroyed by failure of one or two crops. Moreover, there are fewer opportunities for mistakes, accidents, or bad luck to wipe out a crop. Low-input farmers do not have to contend with the consequences of faulty or badly timed chemical applications, either. For these reasons, their farms fare better than conventional farms when times are bad, and the 50-50 odds used in the second set of computer runs are not appropriate.

Therefore, I adjusted the model. The odds were set up so that there would be bad years, in which the income would equal the costs, one-third of the time; fair years, with income valued at one-and-one-half times costs, another third; and good years, in which income would be double the costs, the final third. This model is roughly consistent with the experience of Amish farmers, whose costs rarely, if ever, exceed their income. The results, a doubling of assets in about 18 years, showed a pattern of profit accumulation not unlike that of many conservative small businesses.

Avoiding Ruin

I can think of only two ways for an American farmer to escape gambler's ruin. The first is to leave the game, preferably after a good year, if one comes soon enough. *The second is to diversify, improve marketing strategies, and most importantly, lower costs.*

The soundness of this simple advice is beginning to be widely perceived. In February 1986, *The Wall Street Journal* noted that many New England farms were prospering in the midst of crisis. The article attributed this phenomenon to diversified operations, direct selling of produce in regional urban centers, and lack of expansion in the 1970s when debt-to-asset ratios were rising elsewhere. August Schumacher, Jr., commissioner of agriculture for Massachusetts, observed that the small size of New England farms had kept costs down.

Last August, *The New York Times* reported similar success for the Amish farmers of Lancaster County, Pa. The secret of their prosperity lies in low costs and debts, small farms, and diversification. Their farms typically produce feed corn, alfalfa, hay, wheat, tobacco, vegetables, fruits, honey, dairy cows, beef cattle, poultry, horses, and mules. In addition, they do not think of farm labor as something to be replaced by expensive machinery and chemicals. The Amish approach to life considers the farm labor of the entire family an asset that binds the household and community together.

Most farmers are not Amish. But cost-cutting techniques are available to any farmer who wants to try them. Innovative cultivation and plowing methods can control weeds without herbicides; animal manures and nitrogen-releasing cover crops can be substituted for chemical fertilizers; and integrated pest management can lessen insecticide use. Straw and plant-waste mulches can conserve soil moisture and enhance organic content; crop rotation can reduce pests and disease; and ridge-tilling—growing crops in ridges—can control erosion and improve aeration of the root zone. Profits can also be enhanced through cooperative, regional food-processing facilities, direct sales to nearby urban markets, and the use of specialty crops such as Oriental vegetables, which bring a high return.

DAVID EHRENFELD is professor of biology in the Department of Horticulture and Forestry at Cook College, Rutgers University. He is a member of the Advisory Panel on Biological Diversity of the U.S. Office of Technology Assessment, and edits the journal Conservation Biology.

Reprinted by special permission of *Technology Review,* © 1987.

Your Farm Is Worth More Than Ever

Never mind land prices. Your farm's *internal* resources can let you farm longer and more profitably.

ROBERT RODALE

Agriculture is known to have been practiced for about 10,000 years. If we assume for the purpose of this discussion that agriculture is *exactly* 10,000 years old, then it is accurate to say that farmers and their farms functioned entirely on their internal resources for 9,900 years.

By *internal* resources, I mean the land itself, the sun, air, rainfall, plants, animals, people and all the other physical and human resources that are within the immediate environment of every farm. Historically, farms have been uniquely self-reliant production systems. They have supplied almost everything they have needed from within their own borders.

Think about the land itself, for example. Within the silt portion of all agricultural land are large stores of minerals

	External Inputs To Agriculture *Conventional farming system*	**Internal Resources of Farms** *Regenerative farming system*
THE SUN	A conventional farm captures less solar energy, because monocropping causes the land to be bare of growing plants for a large part of the year.	A regenerative farm makes greater use of the sun's energy, because crop rotation, overseeding, cover crops and the use of fewer row crops keep the land covered with green and growing plants for a greater part of the year.
WATER	Irrigation, which consumes much energy for pumping water, is often used as the primary moisture source.	Rain supplies all or almost all of the moisture, which is readily retained by soils with good organic matter levels and tilth. Plants and cropping systems are selected for moisture economy.
MINERALS	Farms that produce high yields of a narrow range of crops, and that make few or no efforts to recycle available minerals, rely on large imputs of chemically treated sources of phosphorus and potash.	Enough phosphorus and potash to feed crops for years is locked up in most farm soils. Carefully structured farming systems can make those minerals available at economic rates and, by recycling minerals, maintain an adequate level of soil fertility.
NITROGEN	To grow 150-bushel corn, a conventional farmer would pay $56.25 for as much as 225 pounds of nitrogen fertilizer, about half of which would never be used by the crop and pollute groundwater.	To achieve a similar yield, a renegerative farmer supplies nitrogen by plowing down a legume or applying animal manure—at modest out-of-pocket expense. This system also maintains a higher level of soil organic matter which steadily releases nitrogen.
WEED AND PEST CONTROL	Pesticide chemicals brought in from outside the farm—at great cost both in dollars and in damage to the environment—are the prescribed tools of conventional weed and pest control.	Biological weed control methods depend on natural competition for nutrients, water and sunlight—smother crops, canopy shading and crop residue covers, for example—and the abundant and often powerful natural chemicals in crop plants and insects, themselves. Cultivating for weeds is effective. The farmer's knowledge and time are the primary resources.
SOIL-LESS CULTURE	Hydroponics is the ultimate expression and perhaps also the ultimate goal of conventional food production research and planning. Then, virtually *everything* needed to farm becomes an *external* input.	The farm pond, in addition to being a source of water and family recreation, can be used for commercial production of fish and waterfowl.
SOIL		Through a combination of factors, the soil on a regenerative farm has a greatly enhanced ability to provide nutrients, store water and actually suppress crop diseases, weeds and insect pests.
ENERGY	Fossil fuel, mainly from oil and natural gas, is needed in significant amounts to produce synthetic fertilizers and pesticides.	Crop rotation, manure, pest- and disease-resistant crop varieties, solar heating and grain drying, wind-generated power, woodlots and other *internal* resources replace many fossil fuel-based *external* inputs. Methane gas, alcohol and oils for heating and fuel can be produced on the farm.
SEED	Seed produced off the farm has become the norm. Often, new varieties are bred for use as part of a chemical-intensive system. Genetic engineering, which promises to produce even more biologically specialized plants, carries more potential costs and risks than chemicals.	In many cases, a farm can produce its own seed and still maintain genetic diversity and quality. Seed crops for on-farm use and sale can be harvested yearly from the many grasses used for forage and hay, open-pollinated corn and other grains, plus legumes such as clover and vetch.
ANIMALS	The trend toward cash-grain production on more and more farms is eliminating livestock, which previously converted low-quality feed into high-quality protein and provided crop diversification.	Integrated crop-animal systems allow marketing of crops through animals as value-added products. Animals also allow greater crop and management flexibility, and recycle available nutrients.
MARKETING	Many conventional farmers, but especially cash-grain producers, have little control over the prices of their crops. Grain prices, for example, are determined largely by world supplies, economics and politics.	Greater diversity in crops and livestock on a regenerative farm creates more marketing opportunities, especially for seasonal and value-added products that are sold locally or regionally.
MANAGEMENT DECISIONS	Conventional farms are moving more toward systems that require management decisions to be made off the farm. They are relayed to the farmer in the form of "recipes" or packaged systems, legal requirements (for pesticide use), and even computer programs.	When farms are of manageable size, almost all management decisions can be made most effectively by the farmer.

that are frequently unavailable to plants. Those minerals become available little by little, and also circulate within the cropping system of the farm. For thousands of years, that internal resource of the land was relied on to supply all the mineral needs of agricultural plants.

Another internal resource is the air. It is rich in carbon dioxide, which is needed by plants. The air contains 78 percent nitrogen. Although that nitrogen is locked up in a fashion similar to the large supply of minerals in farm soils, some plants and microorganisms have the key to unlock that important resource. Legumes, for example, can provide a good home for certain root-zone bacteria to collect nitrogen from the air and let plants use it.

The energy needs of agriculture are also interesting to examine from a historical perspective. Until fairly recently, agriculture was the primary energy-collecting system of developed societies. The sun was the primary energy source, but agricultural plants and animals were the agents for collecting the sun's rays and embodying them in a form that could be used throughout the year to satisfy people's own energy needs. The farm and the sun together were, therefore, a uniquely economical energy source.

Moisture from rain is also an internal resource of a farm. So are the plants grown, the domesticated animals, and the information about farming itself in the farmer's head. Taken together, all of those important productive elements have — over the last 10,000 years — provided a firm base upon which people were able to build a civilized way of life.

Discoveries Bring Change

But about 100 years ago, agriculture began to move beyond its vast and useful *internal* resources into a production system based also on *external* inputs like fertilizer, pesticides and fossil fuels. What caused this change? Was it the pressure of increasing population? Or did advancing science open doors to an awareness of agricultural processes that, in turn, led to the development of *external* input production?

I think it was both. Even 200 years ago, there was concern that population was expanding beyond the potential of agriculture as it was then practiced to feed everyone. Jethro Tull, the inventor of row cropping and thorough tillage and weeding, warned in the early 1700s that unless farmers adopted his advanced methods, people would either starve or start eating each other.

And about 100 years ago, the great German chemist Justus von Leibig formulated his law of the minimum, which states that crop yield is limited by the level of the major nutrients present in the smallest amount. That insight, and other contributions of von Leibig, opened the way for the creation of the modern fertilizer industry.

Who else contributed to the idea and the technique of increasing crop production by bringing in materials from outside the farm? That would be a long list, indeed. On it would be many chemists, plant breeders, soil experts, engineers and developers of oil and gas resources.

Rather than try to list them, I will present here in a short form the result of their work — a comparison between the major *internal* resources of agriculture and the corresponding *external* inputs. Keep in mind, though, that the information in the two columns on the opposite page is presented as a comparison between the basic elements of a regenerative farming system, which relies largely on *internal* resources, and a conventional system, which is more dependent on the continual use of *external* inputs.

Put Your Farm To Work

One hundred years is a short interval in the 10,000-year history of agriculture. Yet in that time, the production of food for all the peoples of the world has switched dramatically from total reliance on the *internal* resources of farms to systems that are dependent on large amounts of *external* inputs. While that change has enabled farms in the United States and other developed countries to produce ever larger amounts of food, I contend that it has distracted our attention from the enormous value and, indeed, the primacy of the *internal* resources of agriculture.

And the systems using *external* inputs have done far more than merely cloud our vision of the value of *internal* resources. The *external* inputs, themselves, often reduce the usefulness of a farm's *internal* resources.

Nitrogen is one of the best examples. When *external* sources of nitrogen are put into the soil, the rich population of microbes that capture nitrogen from the air does its work less effectively. Often, it ceases to function. And likewise, soil minerals become available more slowly and are recycled much less effectively when input-intensive farming systems are used. And irrigation systems can make rainfall and other environmental constraints irrelevant to crop production, at least in the short run.

Those of us now advocating regenerative farming systems contend that the pendulum should start swinging the other way. We are not against the use of all external inputs in agriculture. Far from it! What we are saying is that farms have tremendous regenerative capacity that can be expanded and put to good use as farmers learn to use their internal resources more effectively.

Yes, your farm is worth more than ever. The title of this article is not an exaggeration. But for many farmers — especially those now following the high-input approach — the tremendous worth of their farms will remain obscured until they are able to base some of their agricultural efforts primarily on the tremendously powerful *internal* resources of their farms.

Part II
The Practice

Chapter 2
Getting Started, Or Look Before You Leap

Farming—A Matter Of Dollars And Sense

JUDY YAEGER

LAWTON, Mich.—Maybe I don't understand economics. (It was my least favorite course in college.) Maybe I don't understand farming. (We've only been working our place full-time for 17 years.) For certain, I don't understand a current buzzword in agriculture—"cash flow"—or why it is used as an index of a farmer's successfulness.

Now I'll grant you that we're not farming for the future. We're not trying to accumulate a large farm estate to pass on to our son (who'll probably end up a truck driver, anyway). We're not shooting for a write-up in *Successful Farming*. We're simply trying to make a reasonable living doing something we enjoy—raising food.

Keeping that goal in mind, I've developed a rather simple approach to the economics of our farm. Each year, I figure up what we sold, and subtract from that amount the cost of growing the crops. Then I use the difference, which I have learned is called "net income," in deciding whether I have made enough money to live on until we have more crops to sell. If the "net" is high, maybe we can afford a vacation. If it is low, we tighten our belts over the winter. If, heaven forbid, it ever turns up negative, we figure we've done pretty poorly, and may even have to pick up a job off the farm for awhile until things straighten out.

Again, in my naivete, I figure there are two ways of getting that important figure, "net," to look promising. One way is to increase sales (called "gross"); the other is to cut down production costs (let's call them "expenses"). Either way works, in my way of thinking. I figure that if we raise $12,000 worth of crops for $2,000, we'll be just as well off as if we raise $24,000 worth with $14,000 in expenses, and a heck of a lot better off than if we raise a $100,000 crop that cost us $100,001.

But there must be some flaw in my reasoning. None of the successful farmers around here, where the measure of success is the size of debt load a farmer can carry, talk much about "net." Mention "cash flow," however, and they're off and running. I've further noticed that the phrase "cash flow" seems invariably linked to the word "problem," as in "cash flow problem."

The other day, I stopped at the local cafe for a cup of coffee and, hoping to learn more about "cash flow," I sat next to a table of farmers, deep in conversation.

"By gosh, my farms (note the plural) grossed $350,000 last year, a real good year. I still have 20,000 bushels of corn in storage. I just hope I can borrow my operating capital at a reasonable interest rate from PCA this year, or I'm going to have a real cash flow problem," said Dick, hunkered over his cup of coffee.

"Yeah, I know what you mean," replied Tom. "I did pretty good this year, too. Sold over $200,000 worth of grain, but my tractor just broke down and I don't know how the heck I'm going to pay to get it fixed in time for planting. Talk about your cash flow problems!" He stirred his coffee worriedly.

Noticing me sitting nearby, and wanting to show that he was not a male chauvinist, Dick nodded to me and asked casually, "How'd you folks make out this year?"

"Oh, pretty good," I mumbled.

"Where're you getting your operating money from this year?" asked Tom, eagerly sniffing about for a bargain interest rate.

Embarrassed, I muttered, "Well, we, ah, always set money aside in the fall to cover our spring costs."

Silence. Stares. Finally, Tom asked, "Then you mean you don't have a cash flow problem?"

Red-faced, I shook my head no. As if on signal, the fellows hitched their chairs back around, turned toward the center of their table and subtly excluded me from further farm talk.

As I got up to leave, I overheard Harry, the youngest of the group but the most successful (he owes almost half a million, some say), begin to speak. "I was talking to my accountant the other day," he said, "and I've decided to get me some hogs. I'll probably wind up with a net loss on the darn things, but they'll really improve my cash flow problem . . ."

Judy Yaeger farms 90 acres near Lawton, Mich., with her husband, John. They raise about 15 acres of vegetables, have 40 acres in pasture and keep 50 ewes.

Know Your Cash Flow In Advance

How much money are you going to get from your blueberries? How much is it going to cost to produce that crop? The answers to those and even more important questions—How much money do I need to borrow to get started? What's my profit?—can quickly be put into the big picture with a basic cash flow plan. This essential budgeting and management tool is used, in one form or another, by all successful businessmen, everywhere.

Basically, a cash flow plan consists of a simple chart that lists all the months of the year across the top (see chart). There is a long, ruled column under the name of each month. Down the left side of the chart is a list of all the commodities and services you will be selling on your farm, plus all other sources of farm income like payment of the annual membership fees for your Clientele Membership Club. On the horizontal line for each commodity, block out the weeks or months when that item will be available and producing income. For example, blackberries and raspberries may be available for only a few weeks in early summer, while enterprises like honey and bee products or rabbits and their products will be generating sales 12 months out of the year.

Sample Cash Flow Chart For Diversified Farms

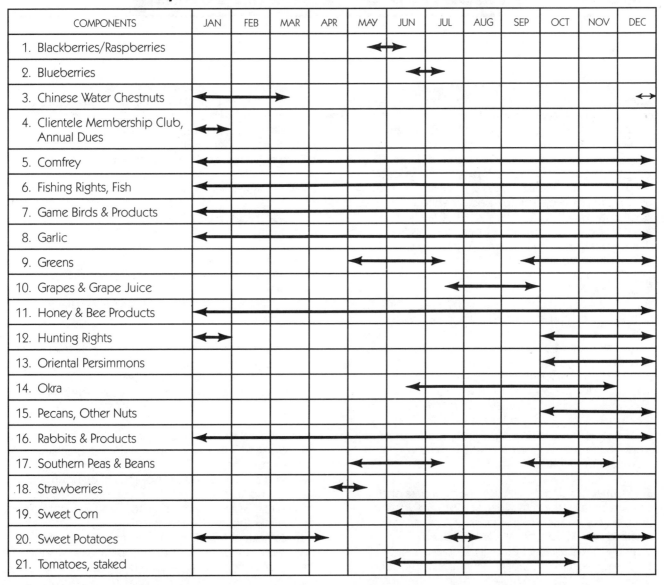

In estimating income and expenses, it is always best to err on the side of caution. Don't stick your financial neck out too far, in other words. After you've determined the amount of money that an enterprise can realistically be expected to bring in, subtract 5 percent, just to be on the safe side. The same rule of thumb applies to estimating expenses, only in reverse. After you've come up with a realistic figure on the cost of producing a crop, add 5 percent. (Expense and income figures can be taken from the enterprise component budgets used in developing your comprehensive farm plan.) That, coupled with the conservative income estimate, will give you a 10-percent margin of safety. That's good insurance, because in farming, as with most things in life, it's not unheard of to end up making a little less than you wanted to, while spending a little more than you wanted to. But then there are the times you'll earn more than you expected, or spend less than you'd budgeted. Or both! Imagine how good that will make you feel. Everybody likes to spend less—and earn more.

Let Your Customers Do The Harvesting

Allowing customers to harvest their own produce is nothing new. Fruit and vegetable growers have long relied on "clean-up operations" to glean what was left after mechanical picking or to extend the harvest after it was no longer profitable to hire pickers for wholesale markets.

But in the last decade, "U-pick," "pick-your-own" and "PYO" have become buzzwords as this method of selling produce has developed into a popular, specialized technique in its own right. An estimated 5,000 farms now sell at least one PYO crop and 96 percent of all the strawberries grown in Illinois are sold this way. One of the gurus of this movement is University of Illinois horticulturist J.W. "Bill" Courter, who has scrutinized successful U-pick farms all over North America—including a farm near Chicago and Milwaukee that sometimes sells 100 acres of strawberries in one weekend, an orchard outside Toronto that needs a 900-car parking lot to handle the crowd, and a group of Indiana growers who cooperatively buy full-page newspaper ads to entice customers from nearby Louisville, Ky. After years of studying the PYO trend, Courter still seems a bit amazed by the phenomenon. It doesn't quite seem logical. "Pick-your-own customers will work harder than hired help; they'll buy produce you couldn't sell at a stand, and they'll drive 30 or 40 miles to do it, sometimes as often as once a week," Courter told farmers at a New England vegetable growers convention.

Even farmers used to dealing with the public are often overwhelmed by the enthusiasm of a PYO crowd. When Richard Tice, who sells orchard fruits and produce through his stand near Woodcliffe Lake, N.J., decided to sell pick-your-own peaches, he was amazed at the response to a few ads in local newspapers. "People really mobbed us; it was unbelievable," he recalled at a New Jersey roadside market conference. "The people came from New York City, from New York state, and gosh knows where . . . They picked green peaches—anything on the trees. Then they would go through the orchard and see better ones. They would dump the green ones out. Then it came to the point where the people couldn't find the better peaches; they would come along and pick the peaches off the ground. There wasn't a thing that went to waste in that orchard. We could have sold the stones off the ground, and there are a lot of them in Bergen County."

Why would farmers invite such chaos? For many, the advantages outweigh the problems. University of Delaware marketing specialists list these advantages and disadvantages:

Advantages

- Need for seasonal harvest labor is reduced.
- Grading, packing and storage costs are eliminated.
- Container costs are lower.
- The grower is paid when the crop leaves the farm.
- No middlemen commissions are collected.
- No labor camps or housing are needed.
- When crops are picked fully ripe, yields sometimes are increased 10 percent to 15 percent.
- Usually, no packing house is needed.
- Grading losses are reduced.
- Grower has more control over prices.

Disadvantages:

- Accident liability increases.
- Managing sales adds long hours of work, especially on weekends.
- Bad weather—or even the threat of it—scares away all those volunteer pickers, leaving you sitting on a crop that can't be harvested.
- Customers may be hard to control and sometimes disagreeable.

If pick-your-own sounds appealing, you'll want to consider two things: how your cropping methods might have to be changed to fit this market, and the techniques of crowd control needed to keep all those amateur harvesters working without being destructive.

Choosing what to plant is your first decision. Fruits seem to do better than vegetables, with strawberries being the first choice among growers. Early berry crops ripen before hot weather discourages pickers. Apples and pumpkins sell well on crisp fall days, too. You'll need skillful salesmanship to convince urbanites that it's fun to spade new potatoes in midsummer, though. The fact that many vegetables ripen in hot weather may be one of the reasons PYO vegetable farms aren't numerous. Another consideration is whether the potential crop is easily picked by inexperienced customers. Some sweet corn is sold on U-pick farms, but pickers often damage immature ears by pulling back the husks to see if the corn is ripe.

Don't be afraid to experiment with variety and off-beat crops, though. Robert Hallock of New Egypt, N.J., sells 17 acres of strawberries, as well as 5 acres of peas, 6 acres of snap beans and 5 acres of an assortment that includes lima beans, peppers, cucumbers and squash—all PYO. He even sells PYO onions, digging them and letting them dry out on top of the beds before making them available to customers. Asparagus has been a popular PYO crop in Minnesota; okra has sold well in the South; and cut flowers in the East. Byone Yoo, an enterprising Korean immigrant who lives near Washington, D.C., contracts with growers near Chicago, New York and Washington to grow PYO Oriental vegetables. Yoo found customers for such crops as bok choy, lobok and Chinese cabbage by picking 15,000 Oriental names out of phone books and mailing them advertising.

How much should you plant? Many farmers recommend a modest beginning. "As a former dairy farmer, I wasn't sure just how PYO would work for me," says Herbert Hoopes of

Forest Hills, Md. "That's why I started out with only a half-acre. I wanted to feel my way. Also, it was my first time growing strawberries. I quickly sold out what I had that first year, but I made a decision to expand gradually. It took me 12 years to reach 7 acres. But I wouldn't have it any other way."

Bill Courter's studies of Illinois strawberry farms show that about 400 customers are needed to clear an acre of strawberries with a 10,000-pound yield. Illinois pickers have been buying an average of a little more than 20 pounds of berries on each visit to a farm. About 75 percent of those customers — the ones in a farm's "primary trade area" — live within 20 to 25 miles of the grower. August Schumacher, Massachusetts agriculture commissioner, believes that the size of a population living within a grower's trade area can limit the practicality of switching to PYO. "Farms 30 miles from substantial towns (30,000 people) or 50 miles from cities of 100,000 may have difficulty attracting customers," he points out.

Planting techniques need to be tailored to PYO sales, too. Planting in succession or choosing varieties with different dates of maturity will extend the season, a consideration just as important to PYO farms as to growers who sell through roadside stands and farmers markets. Spacing is important, too. Hallock planted some of his rows too close when he started out, a common mistake among new pick-your-own operators. Pickers need plenty of room to move around without damaging crops any more than necessary. Many PYO orchards are planting dwarf varieties for the ease and safety of picking.

Crowd Control

As your PYO operation grows, your ability to manage crowds can make or break the business. This involves the physical control of the flow of people onto and away from the farm.

Schumacher divides the task of crowd management into three "management centers" staffed either by hired help or family members:

- Parking lot.
- Fields being picked.
- Checkout stand.

A safe parking spot for customers is essential. If you expect more than 100 or 200 pickers, you'll probably have to hire someone to direct traffic. Jon Nourse, a Westboro, Mass., strawberry grower, tries to keep the parking area within 100 feet of the field to be picked, since many of his clients are elderly. For parking, he uses rocky land that he doesn't cultivate. Clearly written signs direct customers to picking areas.

Like most growers, Nourse hires a field supervisor to hand out containers for picking, weigh containers brought by pickers, point out the best spots for picking and give tips on how to pick. A good field supervisor is a valuable employee, indeed. Nearly 90 percent of all Illinois PYO strawberry operations hire supervisors. They help customers learn to pick efficiently and without causing damage. They also keep people out of fields not ready for picking and away from farm buildings and dangerous machinery. (Locking up such equipment is another wise precaution.) Of course, you'll want to hire someone who is tactful and who enjoys working with people. Many farmers hire retired neighbors for these jobs. Courter knows of one grower who hires local beauty queens.

Checkout procedures vary greatly among PYO farms. F.W. McConnel of Woxall, Pa., has one of the simplest. His farm is at the end of a long, narrow lane. He just meets the customers as they drive up to the house and provides them with containers if they don't have their own. When customers come out of the field, McConnel checks the harvest and collects his money, charging for any containers he furnished. "I want to keep everything as simple as possible," he says. "For me, the less formality, the better. Fewer headaches."

Other growers have found that a simple system eventually brought its own headaches. If they sold by the pint, quart, or peck, a few customers would try to pack more than the normal amount into their containers. Rather than argue with customers over what amount was fair, some growers have decided to sell by weight. It takes more time, but is always fair to both the consumer and the farmer. Many growers also have customers go through the checkout before returning to their cars, since a few dishonest clients inevitably try to sneak out extra containers in the trunk of their cars if the farmer has a drive-through checkout system.

Salesmanship

About half of all PYO customers are repeats from one year to the next, so it's important to do everything you can to get those pickers to return. Providing drinking water, a few rented "portajohns" and maybe some shade at the edge of fields is a minimum requirement for customers' comfort. Some farmers offer a picnic area, although others think this creates litter and trash collection problems. Knowing that many PYO customers come to a farm in search of Old MacDonald as much as to buy produce, some growers have gone to great lengths to provide a rural atmosphere. They provide small livestock "petting zoos" for children. They may haul pickers to fields with horse-drawn wagons. And, if a new building is put up for checkout or sales of other produce, they make sure that it looks rustic. Some farmers also make a little extra money by providing a soft drink machine or selling cider and other refreshments.

Maintaining communications with customers is important. Have them sign a guest register, or put their name and address on a weighing slip, receipt, or postcard. Then, the following year, you can mail them a notice a few weeks before harvest time, telling when your crops will be ready to pick.

Some farmers use a recorded message of prices and crop conditions on an answering machine hooked up to their business telephone. It's best not to tie up your personal phone with such calls. Herbert Hoopes, for example, averages about 225 calls a day during the picking season. Once you've gotten customers to come to your farm, you can encourage them to make larger purchases by giving away shopping bags, information about freezing and storage of produce, and recipes for jams and preserves.

Advertising is essential to attract new customers. About a fourth of the PYO strawberry customers in Illinois are attracted to farmers through ads. Massachusetts grower Jon Hourse uses preseason newspaper ads to let customers know what he'll have available. Then, 10 days before harvest, he places small ads in newspapers and puts up posters in local stores and other public places. He distributes 5,000 brochures annually. He also uses radio for last-minute ads during the peak of the season. A good advertisement should list the produce available and the hours your farm is open for picking, and provide directions to the farm.

The Economics Of PYO

Selling pick-your-own produce can save you many harvesting and marketing costs, and it can be profitable. But many growers underestimate hidden expenses of PYO, and Courter frequently urges them not to underprice their crops.

Courter has found wide variation in PYO prices. Growers near cities seem to be able to charge more than those in rural areas where the average income is lower and residents are more conscious of prices. Retired farm marketing specialist Don Cunnion surveyed Pennsylvania growers and found that some "charge half the retail price; others, slightly over the wholesale market price or slightly under that price (on the grounds that they had no harvest labor involved). The Hallocks of New Jersey say PYO prices in their area usually are higher than the wholesale market.

In arriving at a price, you shouldn't ignore marketing costs. You'll want to add the cost of promotion and hiring field supervisors and other employees to your normal crop production expenses. Jon Nourse estimates that PYO expenses work out to about 5 to 8 cents per quart of strawberries when it would cost him 15 to 20 cents per quart if he hired pickers. Any time the public is allowed on farms, some costs are incurred in the form of soil compaction where fields are used for parking, damage to orchards and crops, trash collection, etc. The American Society for Living Historical Farms estimates such costs at $1 to $2 per visitor. Some PYO farmers recover this cost by charging an admission fee or requiring a minimum purchase.

Other farmers doubt that PYO produce can be sold for less than wholesale prices. As orchardist William Haines Jr. of Masonville, N.J., told a New Jersey roadside market conference: "Our customers were taken to the orchards on wagons. The problems you have in cost, here, is that, many times, you will bring in 30 people on the wagon and 10 or 12 baskets of fruit. This is quite expensive. On a normal weekend in the apple season, we would pick 25 tons of apples and bring them in on the basis of 10 to 20 half-bushel baskets on a wagon. So it takes a lot of wagon rides, a lot of tractors and a lot of gasoline to do this . . . On a pick-your-own basis . . . including customer services and advertising, you have to charge about the same price per pound that would be received for fancy fruit on the wholesale market."

One of the biggest financial risks in PYO sales is weather. Schumacher has found that about three-fourths of all U-pick business falls on a few early summer and fall weekends. Rain can keep customers away during part of those weekends, pushing nearly all of the sales into eight to 12 "peak days." For that reason, few farmers rely only on PYO to sell their produce. "It is this peaking that results in U-pick operations being mostly a supplementary side of a farmer's operation," Schumacher explains.

PYO sales have other benefits besides savings in labor costs that make its risks worthwhile. It can increase sales at your roadside stand, for example. Richard Tice, the New Jersey farmer who was mobbed by New Yorkers in his first attempt at selling PYO peaches, was surprised by booming sales at his stand. "We thought this would hurt our retail operation. But, just the reverse was true," he recalls. "It surprised us. We did an unusually large business the days that this pick-your-own was in business. Some of the people couldn't be bothered picking their own peaches and ended up buying them in the market. They bought other items, so that those days were unusually busy."

Sometimes, too, PYO sales do exactly what every farmer hopes they'll do — increase profits. Charles Logg Jr. a Homestead, Fla., vegetable grower, recounts one experience with his okra crop: "I started out on a $12 (per bushel) market and was picking it commercially. When the market broke, I went U-pick and made three times as much money on the half-field where I had U-pick as I did when the crew picked. I grossed as much on the part that was commercially picked, but by the time returns were figured after subtracting labor for picking, the package, commission, etc., I ended up a lot better off with the U-pick operation."

Handling Money Safely

GEORGE McCONNELL
McConnell Berry Farm
Mt. Vernon, Ohio

Handling money safely is a constant challenge. We think that in our U-pick and farm market operation it is a mistake to have too much money on hand at any one time. That's why we:

- Remove all checks and bills larger than $10 from cash boxes. We often get 100 or more checks a day, as well as an occasional $50 or $100 bill.
- Change money aprons at regular intervals, every hour or more often on busy days.
- "Deposit" money in steel strong boxes welded to the frame of each stand. Only the stand manager has a key to the padlock on each box.
- Maintain constant communication with our blueberry checkout stand via CB radio, since the stand is nearly a half-mile from the house and not in clear sight. We also have CB radios in each farm truck. Two men working around the farm carry walkie-talkies. All radios are turned on all day, every day. They have a range of about 20 miles. With as many as 1,000 customers on the farm on busy days, we find that people and supplies frequently have to be moved on short notice. Our radios really get a workout!
- Use the night depository at the bank daily during busy times to get money in a safe place.
- Advise all employees who handle money not to resist if someone tries to rob them. Chances are good that a hold-up attempt would be heard over our radio network.

Wamp's PYO Operation

A gentleman by the name of Okie Wampuskitty has been charged with the crime of creating a public nuisance. Although his story is fictional, it points to some of the problems that can develop for inexperienced PYO operators. It will be quite obvious that Mr. Wampuskitty was not aware of the management procedures that he needed to operate his PYO operation. Adequate planning is essential, and lack of planning and inexperience are the causes of most of Wampuskitty's problems. The following is an excerpt from the testimony at his imagined trial:

Defense Attorney: State your name for the record.

Accused: Okie Wampuskitty.

Defense Attorney: Is that your real name?

Accused: Yes, sir.

Defense Attorney: People around these parts call you Wamp, though, don't they?

Accused: Either Okie or Wamp. Mostly Wamp.

Defense Attorney: Where do you reside?

Accused: Northeast of Muskogee.

Defense Attorney: Muskogee, Oklahoma?

Accused: Yes, sir.

Defense Attorney: Now Wamp, you've been charged with a crime and you've heard the witnesses against you and you deny the charges, do you not?

Accused: Well, I guess I clogged up the road in front of my place, but I didn't mean to cause nobody trouble.

Defense Attorney: Tell the jury what happened.

Accused: My wife and I decided to try to raise strawberries. We were looking forward to the harvest period, which lasted about three weeks. I was going to hire anybody who would work during this three-week period. The harvest was supposed to produce most of our annual income.

My wife was pregnant and was having trouble carrying the baby. She was to rest as much as possible. I knew that I would be having trouble trying to harvest without her.

With my wife down most of the time, I was having to do a lot of the housework like cooking, getting the kids ready for school, and so on. Anyway, I didn't seem to have time to find hands to harvest the strawberries. So I decided to place an ad in the big city newspaper to let the people know they could come to my place and pick their own strawberries. It was the only way I could figure to get the strawberries picked.

I decided to start on Saturday, because my older kids would be there to help me. The strawberries would be ready for the first picking on that Saturday. My strawberries were ready about a week before the Arkansas berries, for some reason. I ran the ad for a week before that Saturday, because I wanted to be sure all the strawberries were picked.

Saturday morning came. I was eating breakfast when a car horn honked. I looked outside and saw a car with three ladies in it wanting to know if this was the place where you could pick strawberries. I said yes and went out to show them where to go. When I left the house, I never got to come back in until after dark.

The ladies left to pick strawberries, and they were followed by two more cars. These were followed by five more cars. Then I lost count. You won't believe how many cars came that day. My yard filled up right off. People were parking on both sides of the highway, which passes by my place. The cars stacked up for a mile in both directions. There was room for only one car to go down the road, and then the middle got blocked. People couldn't get to my place, and those who were there couldn't get out. Some of the people became mad. My neighbors were mad because they couldn't get in or out of their places. My wife was mad because there was a long line waiting to use our bathroom. I finally told the men and boys to go to the bathroom behind the chicken house and the ladies to go inside the barn. A lady with a French accent told me she was not going to use a pissoir. I was afraid to ask her what she meant. People were lined up to use my telephone. Either one of my customers or my neighbors called the sheriff. He couldn't find the owners of the cars. A lot of them were in my strawberry field.

A lot of people brought their dogs. One of the little female dogs was in heat. The other dogs were chasing that little female or fighting. Things got out of hand — way out of hand. The sheriff went back to town and got a portable public address system. When he returned, he cut my fence and crossed over into my fields so he could come down to the house in his car. When I explained to him what was going on, he couldn't believe it. He got on his public address system and told everybody to return to their cars. Those people with their containers full of strawberries started for their cars. I tried to get the sheriff to announce that the people should pay for the strawberries before they left. He told me to quit bothering him, because he was going to unclog the road. The ladies who had just arrived protested having to leave with their empty containers. The sheriff didn't listen to their protests.

There were quite a few who would not use the chicken house or the barn for a bathroom, and there was a line to our bathroom. One little boy peed on our new rug, and I'm still hearing about that. The sheriff couldn't budge those in the bathroom line who appeared to be in pain.

One by one, the cars began to move out. The only problem was that cars continued to come from the big city. The sheriff had to place two deputies on the highway leading to my place telling them to go home. Those people were mad. I told the sheriff that the next day might be about the same or worse, since it would be Sunday and the city folks might be out in great numbers. He told me that I had better leave the fence down that he had cut and let the customers park in the field. I did. The sheriff was kind enough to station one deputy at my place the next day to direct traffic, but I was told to put up signs, because this would be the only time the deputy would be stationed there. The sheriff told me he was filing a charge against me for causing that big mess.

People are still talking about that day. I hope most of the folks are over their mad spell. One lady isn't, because she wrote me a letter threatening to sue me, claiming I got her water system out of order because there were no bathroom facilities that day. I hope the poor thing doesn't get the word she's pregnant. No doubt, she would point her finger at me, since I have been blamed for everything that happened that day.

After this first day, I began making some money.

Defense Attorney: Pass the witness.

Court: You may cross-examine.

Prosecutor: No questions.

Defense Attorney: Your honor, the defense has no other witnesses except Mr. Wampuskitty's wife, and her testimony would be cumulative. The defense rests.

Court: Rebuttal witnesses?

Prosecutor: None, your honor.

Court: The court will be in recess until 3 o'clock.

The court reconvenes at 3 o'clock.

Court: Ladies and gentlemen of the jury, I have concluded that, as a matter of law, there is insufficient evidence on which to base a conviction. You are therefore excused from further jury duty until tomorrow at 9 a.m. I want to thank each of you for your attentiveness during the trial. Please leave the courtroom quickly and quietly.

Court: Bailiff, assist the members of the jury with their exit.

(The members of the jury departed.)

Court: Gentlemen, I have concluded that based upon the testimony heard against Mr. Wampuskitty, as a matter of law, there is insufficient evidence to support a conviction and I am, therefore, dismissing the charges. If Mr. Wampuskitty is guilty of anything, it is of mismanagement and, of course, that is not a crime. Mr. Wampuskitty, you are free to go. The court is adjourned until 9 a.m. tomorrow.

Analysis Of Wamp's Operation

Although Wamp's operation and trial are fictional, his initial experience as a PYO operator is not far from that of anyone who is just beginning, especially for strawberry and sweet corn growers. The first day can be very hectic.

There are some good features about Wamp's operation. He evidently had good technical knowledge of how to grow strawberries. He probably had irrigation, or else there was just the proper amount of rain at the right time and there were no late freezes. One thing certainly was going for him: his location. He must have had a strategic location to draw such crowds.

Many of the unfortunate happenings at Wamp's place could have been avoided if he had planned ahead. First, his advertising was so effective that it brought too many customers on the weekend. Strawberries will deteriorate fast if they are not harvested when ripe. Wamp may not have had the option of starting his picking before the weekend, but if that option is available, it is better to start on a weekday. The crowds then will not be nearly as large. If Wamp had no option in the matter, he should not have advertised for so long during the week before the harvest was to start.

When you do start the harvest, it is absolutely necessary to have a parking area that can take overflow cars. There was no way that Wamp could accommodate the cars that came to his place. He also had no toilet facilities, which are a must for an operation of any size. We cannot know about his check-out procedure, because, as it developed, he did not have the opportunity to check people out—the sheriff chased everybody off the premises. Undoubtedly, he did not have any insurance for protection against possible lawsuits.

It is not unusual for people to enter the pick-your-own business the way Wamp did, except that there usually is no courtroom activity. Many people decide to let customers come and pick the harvest without any thought of what is involved.

From *Pick Your Own Farming: Cash Crops For Small Acreages,* by Ralph L. Wampler and James E. Motes. Copyright © 1984 by the University of Oklahoma Press.

Chapter 3
How To Build A GUARANTEED MARKET
Establishing Your Clientele Membership Club (CMC)

You can grow good blueberries. Your neighbor can grow good strawberries. The guy down the road can grow good sweet corn. Just about anybody can grow whatever they take a notion to, providing they make sure the crop has enough water and nutrients, and pay fairly regular attention to it.

So what's the big deal about diversification? Why should you knock yourself out to grow only high-value crops like berries, fruits and lambs? Why not just raise corn, soybeans and hogs, like most everyone else?

Because you're not just anybody. As the Rev. Jesse Jackson would put it, "You are — SOMEbody!" You are very special. And so is your farm. Don't you ever forget that. You and your farm are special, because you have one thing the other guys don't have: a market. Not just any old market, but your very own market. Your GUARANTEED MARKET.

Old Ways Aren't Working

Now take a close look at your neighbor who is growing strawberries. What does he do when the berries are ripe? He does what most people do. He puts a classified ad in the newspaper. "STRAWBERRIES. U-pick. 60 cents a pound." Maybe people read that ad, and maybe they don't. Maybe they drive out to his farm to pick, and maybe they don't. Maybe your neighbor will sell his berries, and maybe he won't. Odds are he won't, at least not as many of them as he should have.

"Well, sure," you say. "That's Bill for you. But Jake with the sweet corn, he's going to do alright. Everybody likes sweet corn."

Of course they do, but Jake is even less of a businessman than Bill is. Jake thinks that sweet corn is a sure thing. And since he lives on a paved, well-traveled road, he has no doubt that folks will just come flocking to his place. All he does is put a little sign out by the road that says "Sweet Corn." His sign is an old 1-by-3 that was painted white years ago. He used a stencil to make the lettering, but he thinned the old blue house paint too much. Some of the letters ran a little.

But that's not the worst part of it. Jake may be on a main road, but people would practically have to be starving to stop at his place. It looks like a junkyard, at least what you can see of it does. Jakes's house and fields are way back from the road. The underbrush in the fencerow along the road is every bit of 20 feet high. It has just about grown together over his driveway. All passing motorists can see is this ominous-looking jungle growing over some rusty, old farm equipment on the right side of the driveway and the ruins of an old stone barn on the left side. Oh yes, and Jake's lonely little sweet corn sign, if they can take their eyes off the jungle long enough.

I sure hope Bill and Jake weren't planning to pay off any bills with those crops. Why, at the rate they're going, they'll be lucky to cover the cost of planting. That's because they're taking an amateurish approach to a serious business. The only thing that is worse is the common practice of setting produce out by the road in an unattended "honor system" stand where sun, wind and bugs can get at it all day. Yet those are precisely the methods I see farmers using wherever I go in this country.

HOW TO BUILD A GUARANTEED MARKET

Everybody loves a picnic at the farm!

'Take Charge' Of Your Farm

The two biggest problems facing small farmers everywhere are labor and markets. The reason Jake doesn't have time to make his place look at least presentable, let alone inviting, is because he is working a full-time job 30 miles away in town. The reason he is working off the farm is because he can't make enough money farming to pay the bills. Jake is a good farmer. Then why in heaven's name can't he make any money farming? Because he's so busy driving a tractor before and after work that he doesn't have any time left to really think about how he is going to sell everything that he is working himself to death to grow. All he has time to do in the way of marketing is to dab a little paint on that piece of wood and stick it out beside the road, in the middle of poison ivy vines as thick as your arm. Between you and me, I don't think Jake is going to be in farming very much longer.

And that is a crying shame, because it just doesn't have to be that way. There are a whole lot of people out there who are ready, willing and able to buy whatever America's small farmers can produce. All you have to do as a farmer is make it easy for them. You have to guarantee potential customers a steady supply of high-quality fruits, vegetables and other farm products at reasonable prices. The way to do that is by forming what I call a Clientele Membership Club (CMC). The CMC is literally the lifeblood of your diversified farm. It is as important to the success of your marketing efforts as sunlight, water and nitrogen are to your crops. The CMC consists of a group of loyal customers who will consistently drive past all the farmers markets, roadside hucksters and hand-painted sweet corn signs for maybe 40 miles, just to patronize your farm. I recommend that farmers charge an annual CMC membership fee of $25, payable the first of the year. Some farmers in more populated and affluent areas charge $50 a year.

Why would people come only to your farm and pay for the privilege? Because, like I said before, you and your farm are special. You are the new breed of American farmer. You are perceived as a dedicated individual — a family man, religious, respected in your community — who is in the business of providing your customers with wholesome, nutritious and contamination-free fruits, vegetables and a cornucopia of other high-quality farm products. You also offer fresh fish, lamb and nuts, game birds, rabbits and honey. You stable and feed their horses, offer a safe place to go hunting and provide the whole family with the opportunity for recreation in a relaxing rural setting. In short, to borrow an advertising line for a popular seaside resort, your farm is the antidote for the craziness of today's civilization.

Alright now, enough of the philosophy behind a CMC. Let's get down to the nitty-gritty. The psychology of a CMC can wait. What you need to know right now is exactly how to organize a CMC, increase its size to about 1,000 members and make it thrive.

Hand-Picked Prospects

The crucial first step is to hand-pick members for your CMC as carefully as you would select a pint of raspberries for use as a gift. You want to identify and contact people who have money. Now I'm not necessarily talking about the country club set, although 100 or so of those folks would be mighty nice. You want people who have good jobs. But, more importantly, you want people who share your belief that food is one of the secrets to a healthy, happy and long life, not just another raw commodity to be processed, manufactured and warehoused like auto parts.

You just can't beat vine-ripened fruit — right off the vine!

Your customers must also have the same philosophy as the ant. They believe in stocking up for the winter, which means they have to have at least one freezer. The more freezers a family has, the better. After all, where else are they going to keep all of the food you produce?

Where do you find these people? Anywhere and everywhere. Start putting your club together by sending an announcement about your "farm of the future" to your local newspaper. Feature writers, farm editors and even food writers are always looking for a good story. Strategically place posters and brochures about your farm on bulletin boards in office buildings. Give the Chamber of Commerce a call and and ask for names of companies that have a known commitment to promoting good employee health, nutrition and fitness. Visit other places where physically active, health- and nutrition-conscious people are likely to be, places like exercise clubs, private swimming pools, dance studios and health food stores. Don't forget churches and temples, especially those of denominations that encourage healthful eating and keeping a well-stocked pantry. Greeks, Italians, and members of other ethnic groups are apt to be especially interested in your lamb and rabbits. Contact area doctors who specialize in treating people with food and other allergies. And it surely couldn't hurt to drop in on local appliance dealers. Give them a handful of your brochures so that they can give one to everyone who buys a new freezer. The main thing is to actively promote your club, your guaranteed market. If you don't, something terrible happens—NOTHING!

Here, in his own words, is how Richard M. "Rudy" Rudowski of Okemos, Mich., is going about it:

"A local firm called AD-Vantage sends a packet (of no more than 12 ads) about every other month to eight different marketing districts in the Lansing metropolitan area. My current strategy is to make a 3½- by 8½-inch ad to enclose in their next mailing, targeted toward only three districts in the metro area. Each district consists of approximately 10,000 households.

The first blanket mailing to a particular district will go out in late January or early February. Thirty days later, a second mailing will go out, but it will be targeted toward predominately townhouse and duplex neighborhoods that have families with children, but small backyards. If, within 10 days of the January and February mailings, I don't get at least 50 memberships, I will knock on doors within the neighborhoods to sell the clientele club concept, in person.

The March mailing should yield about 100 members, while April's goal is 200, for a total of 400 (includes 100 members from blanket mailing) member families for 10 acres. Note that the first mailing is used to get your name into the public's eye. The follow-up mailings are more focused.

A general rule in sales is that for every 25 contacts, you can generate 12 meetings and three sales, a roughly 8-to-1 ratio of contacts to sales.

High-quality, contamination-free fruits and vegetables... that's what American consumers demand and are willing to pay for.

After the members are recruited, they must be kept informed. Starting March 15, an 8½- by 11-inch newsletter, printed both sides and folded in half, will be mailed to each paid member family. The first issue will describe the farm and again describe how the club works. Later on, this newsletter will be a good way to mention birthdays, anniversaries, etc., because people do like to see their names in print. Issues will be sent every six weeks.

The newsletter is of a chatty nature. The real effort (to get people out to the farm) will be a postcard sent every two weeks announcing what is ready to pick. You can make a master list of all your crops and, for each mailing, just add checks next to those items that will be available over the next two week period.

When setting membership goals, keep in mind that even though mail is an effective advertising medium, you may need to go out and knock on doors to sell yourself and your operation.... Next year, we are planning on increasing our acres planted and, hopefully, relying on word-of-mouth advertising to sell even more memberships. In any case, we will not take advertising for granted. Advertising does not cost... it pays!

This may look expensive (about $2,864 total) and time-consuming, but I feel that ensuring a known market for our produce reduces a lot of stress and worry. I think that, alone, is worth it!"

Laminated identification cards will help make customers feel that they really belong to your farm. Assigning membership numbers can help you track sales by family.

Way to go, Rudy! While his methods may not be just right for your area, Rudy's efforts show you the type of commitment and innovative thinking that must go into establishing your guaranteed market. Good things don't come easy.

Put It In Writing

Once you identify prospective CMC members, you should tell them about all of the commodities you have for sale and briefly explain how your CMC works. Be sure to explain that fishing rights cost $15 extra and that products such as sweet potatoes, honey and game birds are neither PYO items nor do they qualify for the 40-percent discount. Always give people a brochure that lists all of the commodities you grow or sell. The brochure should also state your business hours and contain a clear explanation of just how your CMC works, with emphasis on the fact that your PYO prices are always 40 percent less than than what the supermarket charges for the same commodity. A truly well-done brochure will also include biographical information on your family and farm, and even a photograph of your family with the farm in the background. Don't forget to include a map and detailed directions to your farm.

This whole approach to farming is extremely people-oriented. You have to like people to deal with the public like this, because your entire business is built on solid one-on-one relationships. That's quite a contrast to the usual way of farming, which is highly impersonal and anything but people-oriented. Maybe that is one of the reasons this club concept is more readily understood and appreciated by consumers than it is by many farmers and people from USDA and the land grant colleges. Americans just love to belong to things. We are a nation of joiners. Americans become very possessive of groups they belong to, so don't be surprised if you overhear some of your CMC members talking about "my" farm or "our" farm. In fact, encourage them in that way of thinking. Just imagine what good word-of-mouth advertising that will be when Mrs. Jones tells an envious Mrs. Smith all about the lovely afternoon she had picking blueberries, soaking up the fresh air and sunshine at "our very own little farm." Why, Mrs. Smith will be calling you before the sun sets.

A Sample Letter For Prospective Customers

The personal touch is extremely important in dealing with your customers. It's even more critical in courting prospective customers. As soon as you have talked with someone, either by phone or in person, and determined that they would make a good CMC member, follow up the conversation with a personal letter inviting them to visit your farm.

The following is an example of the kind of letter you should send. (It was submitted by Betty and Jim Beck, who have a farm near Leavenworth, Kan.) This letter could easily be kept on a computer disk and modified or personalized, as needed.

July 19, 1987

Beck Farms
Betty & Jim Beck
Box 3324
Leavenworth, Kan. 66027
(Area code & phone number)

Mr. & Mrs. John Doe
314 Cherry St.
Kansas City, MO 64108

Dear Mr. & Mrs. Doe:

You and your family are cordially invited to visit our farm, which is located on Kansas Route 19, just 15 miles northwest of Kansas City, Mo.

On our 33 acres, Betty and I are in the business of farming to provide families like yours with wholesome, high-quality, nutritious and contamination-free fruits, vegetables, nuts, meat, honey and many other farm products.

Ours is a Pick-Your-Own farm, but it is probably not like any other U-pick farm you ever visited before. That is because not just anyone driving down the highway can stop here and pick whatever they want, whenever they want. Our luscious strawberries, blueberries, sweet corn, grapes, sweet potatoes and other crops are reserved especially for members of our "Clientele Membership Club" (CMC). For an annual fee of just $25, members of our CMC can visit our farm and harvest fresh fruits, vegetables and nuts for just 60 percent of what the same items are selling for in the local supermarket. That's a full 40 percent off the retail price for produce that you know is as fresh and wholesome as it can possibly be, because you just picked it, yourself. We do ask, though, that you provide your own picking-harvesting containers and also containers for carrying your farm-fresh produce home.

In addition, we have a beautiful, 8-acre fishing lake stocked with largemouth bass and blue gills. It provides excellent fishing, year-round. When you join our CMC, you may also purchase fishing rights for your family for just $15 in addition to your regular membership fee of $25.

For the children, we have a play area, complete with (list features such as swing set, sliding board, jungle gym). We also provide restrooms for your comfort and a picnic area to complete your day in the country.

We have the following items for sale on our farm:

1. Strawberries
2. Blackberries/Raspberries
3. Peas and Lima Beans (free shelling provided)
4. Sweet Potatoes

5. Greens (list them)
6. Blueberries
7. Sweet Corn
8. Honey and Honeybee Pollen
9. Game Birds (list them)
10. Nuts (list them)
11. Rabbits
12. Fishing for Bass and Blue Gill, Year-Round
13. Lamb
14. Bedding Plants (list them – you buy wholesale and sell retail)
15. Grapes and Fresh Grape Juice

This is an opportunity not available to everyone. Our farm will provide for the freezer needs of only a few hundred families.

Betty and I would be delighted to have you and your family join our "family" of members, who are now enjoying the best food they ever ate and relaxing family recreation, all in a healthful country setting. We stress the healthful atmosphere of our farm because we use few, if any, synthetic pesticides and fertilizers. Our goal is to provide you with farm-fresh, highly nutritious and contamination-free foodstuffs.

We hope our CMC members share Somerset Maugham's philosophy: "It is a funny thing about life . . . If you refuse to accept anything but the best, you very often get it." Mark Twain expressed our personal philosophy best: "Always do right. This will gratify some people and astonish the rest."

We look forward to hearing from you very soon. Just give us a call or drop us a note to let us know when you'll stop out for a visit. Enclosed is a map with complete directions to our farm. Our hours are (list them).

Thank you for your consideration.

Sincerely,

Betty and Jim Beck
Beck Farms
Leavenworth, Kan.

Monitoring The CMC

Never lose sight of the fact that yours is a diversified farm with a complex mix of overlapping crops and enterprises. That makes it absolutely essential to keep good records, plan realistically and pay close attention to detail. It means you should have a personal computer and use it to store and analyze every scrap of information about your farm. Please don't let the idea of a computer scare you. More and more of them are what is called "user-friendly." They are designed for people like you and me who don't want to memorize a 400-page instruction book and learn a mathematical language that looks as difficult as Chinese. About the friendliest computers around are made by the Apple company. Just by following the "menu" on the display screen, you can do about anything you want to on an Apple. An Apple IIc and an Imagewriter printer would meet your needs quite well, and for only about $1,000. AppleWorks software, which will do everything from handle your mailing list and production records to run spread sheets, will cost another $300 to $400.

Think of a computer as another farm management tool that will let you do your job quicker and better. Computers on the farm are here to stay, because, just as farming during the last 25 years became what a lot of folks call "input-intensive," the farming of the next 25 years will be "information-intensive." The more you know about your total farming operation and the more ways you can analyze that information, the better.

For keeping track of your customers and their purchases, it will be necessary to give each family a code number. This is the same number that will be on the laminated CMC membership card given to each member family or individual member. You will also need to assign a code number to each crop or product you sell. Using those numbers, you can precisely record how often each member visits the farm, the days and dates, and hours of the day they visit, how much of each item they purchase and the price paid. Constantly studying those numbers will allow you to precisely adjust the operation of your farm to give customers just what they want.

You will also want to keep a close eye on your membership "renewal rate," the rate at which customers renew their annual memberships. Yes, there will be people who will not renew their CMC memberships, just as there is constant turnover among members of any other member organization, whether it is the National Rifle Association, the American Automobile Association or even the Farm Bureau Federation. Without years and years of experience to fall back on, no one can say whether you should expect average renewal rates of 60 percent or 95 percent. Your goal should be to renew as many members as is humanly possible. That's because renewal income is almost pure profit. One of the best examples of that is the magazine business. A magazine may spend $8 to get a first-time subscription payment of $10. Printing and mailing the magazine quickly uses up the remaining $2, so the magazine makes little, if anything, on the new subscriber. But renewing that subscriber may cost only $1, which results in a good profit no matter how you look at it. The same holds true for your CMC members. You go out of your way to get people to join the first year. So you definitely need to keep as many of them as you can, for as many years as you can. To help keep your renewal rate strong, you might want to offer "premiums" or "gifts." That's a common technique that consistently gets good results in just about any kind of renewal business. Your premium wouldn't have to be expensive. A small jar or two of honey or jam would probably work just fine. It may cost you a couple of dollars, but it would be well worth it to bring in $25 to $50 each from the majority of your members at the beginning of the year.

Annual Media Days

Have a party and invite the press. I recommend that farmers hold two media days a year. One should be a fish dinner with fresh fish from your pond. This will be at the start of blueberry harvest, so don't forget to serve fresh blueberry pie for dessert. The other media day should be held in early fall when your main dishes would be quail, sweet and sour rabbit and sweet potatoes. You should make every effort to get all of the local media people there, including public television. Since your farm will be no more than 40 miles from a population center of at least 50,000, the local media force will be sizable. There are three reasons to get the press out to your farm: 1) so that area journalists get to know you as an outstanding farmer in the community, 2) to educate journalists about the realities of farming so that they can educate their readers, and 3) to show off your farm. When it comes to local news about farming, journalists tend to contact the same farmers again and again. You want to become one of their most reliable, knowledgeable and quotable sources. You want the news photographers to immediately think of your farm whenever they need good feature photos. And besides, a lot of those folks might just make good CMC members.

HOW TO BUILD A GUARANTEED MARKET

Incorporating The Farm

There is a lot of loose talk about big corporate farms taking over agriculture in this country. The implication is that to incorporate a farm is somehow bad. Nothing could be further from the truth. The small farmer should by all means incorporate the farm. All that involves is having an attorney draw up the articles of incorporation and the corporation's by-laws. (You should already have a working relationship with a good attorney, just as you should with an accountant or other financial advisor. Remember, people really don't plan to fail, they fail to plan. So make sure you have definite legal and financial safeguards for the future, starting with a will.) Name the corporation whatever you want, just as long as you don't use a name that someone else has already registered. The law requires that a corporation have a board of directors. It's a good idea to have your attorney be on the board. Other directors should be family members, who own all of the stock in the corporation. There are many advantages to incorporating your farm, too many to go into here. Your attorney can tell you all about it.

Teach A 'Cook' To Cook
And you'll have a loyal customer for life.

Now I am not a great *"Chef de Cuisine."* I can't even be classified as a cook. The only thing Lottie allows me to prepare in her kitchen is coffee. On the other hand, I am a connoisseur of fine foods and have dined in many of the finer restaurants around the United States and the world. I firmly believe that family meals should be just that, family events. They should be planned, properly prepared and relished, not just thrown together, wolfed down and (burp!) forgotten. Like the French, I believe mealtime should be a festival.

Ideas like that sometimes seem a bit old-fashioned these days. "As the baby boomers reach middle age . . . they will fill their refrigerators with convenience foods, fresh fruit, vegetables and fish. Home delivery of groceries will be popular and they'll dine out more often at ethnic and theme restaurants," reports *Research Alert* (Vol. 4, No. 26) from Media Resources Inc.

Yes, we have reached such a level of sophistication in this country that many young American women and men do not know how to prepare much of anything for the table. You can give most of these young cooks the best cut of meat available—a standing rib roast—and they will turn you against beef!

Fortunately, not everyone in America is living on Big Macs or frozen entrees and side dishes for which they give thanks to Stouffer. The women and men who will join your Clientele Membership Club care a whole lot more about the quality of their food than the average American does. Instead of convenience, they're more interested in the freshness, nutrition and purity of what their families eat. And because they care more about the quality of their food, you can be darn sure they are going to take greater care in storing, preparing and serving the food they buy from you.

That's where the small farmer can really reinforce his clients' commitment to buying farm-fresh products and have a significant and positive influence on the quantities and types of produce they buy, now and well into the future. That's not as hard or mysterious as it sounds. All you have to do is help your clients learn more and better ways to prepare their food. I'm not suggesting that you open a cooking school, although if there is enough customer interest and potential profit in a cooking class or two, it sure might be worth a try. There are a lot of simple, inexpensive things you can do, like having truly superb recipes handy to pass out to your clients.

You can also strengthen your customers' resolve to buy contamination-free foodstuffs from you. "It should come as little or no surprise to the produce industry that about 96 percent of those surveyed in the Fresh Trends 1987 survey are at least somewhat concerned about the presence of chemical residues on fresh produce items," according to FOCUS 1986-87, a special supplement of *The Packer,* the weekly newspaper of the produce industry. That is based on answers by the 1,285 people who responded to a survey mailed to households that are typical of the latest U.S. census data in terms of age, population density, household size and income.

Your customers could probably put that statistic to good use. "Just imagine," one of them might say at a cocktail party, "96 out of every 100 Americans worry at least a little bit about the chemicals in our food." Why, they'd be the hit of the evening, especially with more information on the current sad state of affairs in our national food system. You won't have to look much farther than the local newspaper for a lot of this information, either. For example, in just one week earlier this year, the local newspaper carried articles about illegal chemical residues in imported foods, alarmingly high rates of salmonella contamination in poultry and a newly discovered "toxic fog" found floating over fields treated with common farm chemicals. I'm in no way suggesting that you try to scare people. **Concerns about the quality of our food are nothing to be taken lightly. So don't scare people, inform them—for their own good and yours.**

'Mealtime Magic'

Also, you can use the experiences or customs of your own family to enrich mealtime for your clientele. For example, the small-farm family often has the opportunity of sharing most of its meals together, rather than each member eating off the stove separately. That, in itself, is a blessing in these modern times. Most of your CMC members won't have that same opportunity, but I'm sure they would want to dine like your farm family on the rare occasions when they are all together. So tell your customers about your mealtime rituals.

Once the family has assembled at the table, the father should take charge and say grace. By that example, the children will learn that a blessing is an automatic part of the meal. Soon, they will want to share the honor of saying grace. They'll even want to give their very own blessing. By all means, praise and encourage them in this. You'll soon be amazed at all the things children will think of to be thankful for and how they will express themselves.

In our home, we hold hands while I say grace. After the blessing, I kiss the ladies' hands. This is the way my granddaddy and father did it before me. When my nieces, Sham and Tam, were little girls, they so enjoyed the hand-kissing that they insisted on sitting on each side of me at the table. If Sham was first to have her hand kissed at breakfast, Tam had to be first at lunch, and so it went.

Mealtime is an excellent opportunity to teach children good manners and other social graces, such as the art of good conversation. Topics of conversation should be as varied as the items on the menu. No meal is complete without a well-balanced blend of good food and good talk. Of course, the most important part of the conversation is when each member of the family thanks the cook and lets mother or father know how much the meal was enjoyed.

A Whatley-style farm produces a cornucopia of fruits, vegetables, meats and other foodstuffs that lend themselves to many fabulous and delicious dishes. Consider your favorite recipes "trade secrets" that you should pass along to your clients, to increase both their dining pleasure and your total sales. For example, when the sugar snap peas are ready for picking, be sure to give each family instructions for freezing the peas, and a recipe they can use for dinner that night. Ever think about keeping a few cookbooks on hand for sale to your customers? How about writing your own cookbook? You could give a copy to each CMC member as a birthday, anniversary or Christmas present, and sell the book to other cooks in the area. Just use your imagination!

Insurance—A Liability Policy Is Just Not Enough
VICTOR R. MILLER

Farms in general—and Pick-Your-Own farms in particular—are notorious for their high incidence of insurance claims for property loss and personal injury accidents. There is one other big reason why insurance companies consider PYO farms anything but good risks: You have non-farmers doing farm work!

Your insurance policy is a fragile line of defense against lawsuits. While the insurance coverage on your farm is a positive step to protect your assets against the rising costs of accidents and steadily mounting jury awards in liability cases, it is just not enough.

For example, an insurance policy does not protect you against increased premiums that result from filing several damage or loss claims. Neither does it guarantee that you will always be able to buy coverage. Insurance companies can and do cancel or refuse to renew policies with a history of frequent claims or even greatly increased risk. Just ask your local farm chemical dealer what he has to pay for insurance now, if he can even get insurance. According to several recent industry surveys, more than a few chemical suppliers are operating with no coverage.

Why are insurance companies getting such a case of the jitters? Farm chemicals, whether the client is making them, storing them, having them applied or applying them, himself, are just becoming too risky. The U.S. Environmental Protection Agency can't really say whether many chemicals are "safe." Fertilizers and pesticides are finding their way into a growing number of drinking-water wells around the country. Farmers in Connecticut are having to pay to provide some 600 homeowners with clean drinking water. Property owners are suing over pesticide spray drift incidents. And homeowners are trying to hog-tie the lawn-chemical industry with new legal ropes. Some seven communities have now passed ordinances that require advance notification of spraying and posting of treated lawns with signs that say children and pets should keep off the grass for a day or two after application. Nationally, a group made up largely of mothers with young children is trying to halt the use of chemicals on lawns, parks and other public places. A citizens' group founded in 1986 to oppose the lawn-care industry is called HELP, an acronym for Help Eliminate Lawn Pesticides. You might want to keep all that in mind when planning your spray schedule. Many of your customers might not be as eager to pick blueberries if they feel worried about what's been sprayed on the plants or even between the rows. A word to the wise should be sufficient.

If you are being sued by someone who was injured on your property, your attorney will want to show the court that you took all prudent steps to prevent accidents and to fully inform customers of any risks they might be exposed to on your farm. "Attractive nuisance" is a term that is often used to describe necessary but potentially dangerous objects that just naturally attract people, things like ponds and tractors. Children seem to find such items especially enticing.

To keep from coming out on the losing end of a big damage settlement, you must be able to convince the judge or jury that you have taken prudent steps to prevent accidents on your farm. Your pond, for instance, should be enclosed by a substantial fence with a gate that is kept locked. Ugly as they are, "No Trespassing" or "Keep Out" signs should adorn the fence. Having a life-saving buoy attached to a long rope hanging on a post near the water's edge is a good idea.

All farm equipment such as tractors, wagons, trucks, mowers, sprayers and rakes should be stored well away from any areas frequented by customers. Never leave ignition keys in motor vehicles. All machinery storage areas should be posted with signs that warn people to stay away from the equipment. Never operate mowers or other implements that could throw rocks or other debris anywhere near customers.

Ladders and any other tools used by your customers should be in good working order. Carefully instruct your customers in the proper use of these devices.

Despite all that, accidents can and will happen on your farm. Keep fire extinguishers, first aid kits and other emergency equipment handy and properly maintained. Make sure the local ambulance corps and fire department know how to find your farm, and keep their phone number right beside each telephone.

By following the steps outlined here, you will easily be able to show the court (if, God forbid, it goes that far) that you have taken every reasonable precaution to guard against accidents. Remember, your insurance policy covers you adequately ONLY if you have taken positive steps to prevent accidents. Whether that means regularly controlling weeds, snakes, lizards, ants, flying insect pests, or handling and storing chemicals as cautiously as possible, always "THINK SAFETY."

Victor R. Miller is a "reinsurance" consultant in Monroe, Tenn.

Say It With Signs

Here are some ideas for signs that you may want to use around your PYO farm to help make your customers' day in the country safer and more enjoyable. They may also make your insurance agent smile and lessen the serious legal problems arising over any accidents that occur.

Parents — Do You Know Where Your Children Are?

Not Responsible For Lost Or Stolen Property

Low Maintenance Road — Travel At Your Own Risk

Not Responsible For Valuables Left In Unattended Vehicles

Please Don't Feed The Alligators — NO SWIMMING

SLOW — No Wake!

SLOW — Turtle Crossing

Pick Here!

Members Only

Not Ripe Yet

CAUTION — Low-Flying Ducks

Thank You! Call Again

Have A Nice Day

Play It Safe — Buckle Up *(at exit)*

Off Limits — *(near machine shed or other potential disaster site)*

Chapter 4
Equipping Your Whatley-Style Farm, Sensibly And Economically

Buy what you need to do the job right — and on time.

Equipping a small, highly diversified farm is like setting up any other business: You can spend a lot or a little. How much you spend is entirely up to you. After all, you are the one who is going to have to live with the consequences of your purchasing decisions. Shortcuts taken now, while they may "save" money at the beginning, will result in higher costs and demands on your time, delayed field work, reduced quality and performance and other management headaches a whole lot sooner than you ever imagined. Before you know it, you're right behind the old eight ball, frantically trying to figure out how in the world you're ever going to accomplish even half of the things that need doing. So don't let that happen to you. Be smart. Buy right the first time, and really save.

Those Options Add Up

Fully equipping a Whatley-style farm with all new equipment will run you from about $69,000 to $142,000 at 1987 prices. Granted, that's an awful lot of money. But more than $73,000 of that higher figure is for items that you may not need for six to eight years — or may never even purchase, depending on your location and crop mix.

Optional items include a $40,000 grape harvester, a $3,500 grape crusher/stemmer and $3,000 for stainless steel tanks for storing grape juice. Don't worry about buying all that until you have 10 acres of trellised grapes about to come into full production.

Precision, in-the-row weeder, like this "Weed Badger," is just the thing for controlling weeds without herbicides in vineyards and orchards.

EQUIPPING YOUR WHATLEY-STYLE FARM, SENSIBLY AND ECONOMICALLY

Not everybody is going to want to grow sweet potatoes, either. That makes an $11,000 sweet potato washer/grader optional.

A tree shaker costs $10,000, but you won't need one of those for six to eight years after your pecan or other nut trees have been planted. A shaking service should be part of your leasing agreement with Clientele Membership Club members. They will lease your nut trees on an annual basis for $50 to $60 per tree. The same leasing procedure should also apply to fruit trees and even sugar maples.

Back To Basics

Take that optional equipment out of the picture and you can concentrate on the lower figure of $69,000. Remember, that's the price of new equipment. Buy used equipment, and you can accomplish the exact same thing for a much more manageable $34,500. I'll bet you feel better already, don't you? You just saved $107,500.

Now buying good, used equipment is not as difficult—or risky—as it might sound (see "Bid Better, Buy Cheaper" and "How To Buy A Good Used Tractor" in this chapter). It's a sad commentary on the times and the condition of agriculture in the United States, but there is a tremendous amount of decent farm equipment on the market today. If you take the time to look for them, you'll find some real good buys out there.

Not long ago, I was amazed by the liquidation sale of a 4,000-acre cotton farm in Alabama. They had four cotton pickers, which they needed with 4,000 acres of cotton. Those things cost $100,000 apiece and they couldn't give them away. But that same farm also had 23 pickup and dump trucks. Now what they needed with 23 trucks, I'll never know. Maybe that's part of the reason they went out of business. Not one of those trucks was more than 4 years old! Folks snapped them up for a song and a dance.

There's nothing like a PTO-driven rotary tiller to save time and work up a fine seedbed for vegetables and other crops. This Bush Hog RT series tiller comes in cutting widths of up to 96 inches for 100-hp tractors.

The same thing holds true for things like milk tanks from dairies that are going out of business. Stainless steel milk tanks and dairy refrigeration units are excellent for holding freshly pressed grape juice at optimum conditions. There is a big demand for fresh, contamination-free grape juice, especially from amateur wine-makers. Of course, you could always go into the wine-making business, yourself. More and more small wineries are springing up around the country all the time. Maybe part of the reason for that is the fact that so many people who try making wine at home end up making vinegar, instead.

Whether you decide to include all of the options or just stick with the basics on your farm, here is a list of the equipment you'll need. All prices are 1987 list prices for new equipment.

BASIC STARTUP EQUIPMENT

Tractor, 52-hp	$18,500
Pickup truck, 8-cyl.	11,300
Turning plow, 2-bottom, 16-inch	1,000
Disk harrow, 7.5-foot	1,140
Fertilizer/lime spreader, 6-foot	480
Manure spreader	2,500
Rotary tiller, PTO-driven, 54-inch	1,370
Disk hiller, 2-row	500
Subsoiler	1,000
Transplanter, 1- or 2-row	1,000 to 2,000
Sweet potato digger	2,500
Cultivator, 2-row	900
Mechanical weeder, precision in-the-row	5,500
Planter, 2-row	1,800
2 Sprayers, 50- to 100-gallon each @ $1,000 (One for herbicides, one for insecticides, foliar fertilizers, etc.)	2,000
Air compressor	500
Post hole digger, 9- and 24-inch augers	1,500
Rotary mower, PTO-driven, 5-foot	900
Flail (grooming) mower	1,600
Middle buster, 16-inch	260
Chain saw	450
Front-end loader	4,200
Scraper blade, rear-mounted	600
Mulching machine (Shreds & blows large, round bales)	5,000
Hand tools	1,500

OPTIONAL EQUIPMENT

Grape harvester	40,000
Grape crusher/stemmer	3,500
Snow blower	2,000
Tree shaker	10,000
2 Stainless steel tanks, 1,000-gal. each	3,000
Pea and bean sheller	3,800
Sweet potato washer/grader	11,000

3-Tractor Vegetable Growing System

A guide to tractors and other machinery essential to the efficient operation of a small, diversified farm anywhere in the United States.

TOM MORRIS

To farm successfully, you need a versatile line of machinery that will allow you to perform a wide variety of field operations quickly, efficiently and, most importantly, at just the right time. Field operations will be many and varied, because on a Whatley-style farm, you will be planting at least half a dozen annual crops in addition to taking care of an equal number of permanent plantings. Depending on the demands of your Clientele Membership Club, you could easily find yourself planting upwards of 25 different, but complimentary enterprises at any one time. Those enterprises will range from small-seeded crops like spinach, other greens and carrots to larger-seeded crops like peas, beans and sweet corn, plus transplanted crops like sweet potatoes, broccoli and cauliflower.

With so many diverse crops to plant and care for, having the right machinery ready to go at just the right time is absolutely essential. Having your farm well-equipped not only allows you to get field operations completed on time, but it saves you a lot of money on labor, chemicals and other inputs.

If you're just starting out, the row spacings you use will be determined by the tractors you buy. Available tractors dictate either 5-foot or 6-foot wheel spacings. The narrower, 5-foot spacing makes better use of your land, resulting in slightly higher yields per acre. If your existing machinery line uses the 6-foot spacing or that is the most readily available size of tractor in your area, stick with the 6-foot spacing. Switching to 5-foot spacing would cost too much for new equipment or alterations to your present cultivators, planters and other implements. Standard row spacings for each category of crop are:

Direct-Seeded Crops
1. Small seeds (lettuce, carrots)
   5' spacing — 3 15-inch rows
   6' spacing — 3 18-inch rows
2. Large seeds (beans, sweet corn)
   5' spacing — 2 30-inch rows
   6' spacing — 2 36-inch rows

Transplanted Crops
1. Bare Ground
   5' spacing — 3 15-inch rows or 2 30-inch rows
   6' spacing — 3 18-inch rows or 2 36-inch rows
2. On Black Plastic
   5' spacing — use 4' plastic
   6' spacing — use 5' plastic

With so many crops to plant and handle, it is not economical for the average small-scale grower to mechanize every operation. But you're not the average grower. You don't have many of the same harvesting and handling problems as other farmers, because customers do the picking and haul the produce right home. That frees you to mechanize two operations that traditionally require a lot of hand labor, transplanting and hand weeding. And you can save a lot of money on labor and chemicals when you mechanize weeding.

Before you jump into mechanical weed control, though, look at your entire machinery system, not just what type or style of cultivators to purchase. Cultivation of crops is at the very end of the line of your field operations, and cannot be done properly unless every previous operation has already been performed just right. For that reason, you should strive to set up the most efficient system of production that your money allows. In building your system, always keep in mind an overview of what you want the system to look like when it is complete. Hopefully, fewer mistakes will be made this way. When mistakes do occur or expansion or some other change in your farming makes a piece of equipment obsolete, I believe it is better in the long run to sell the piece of equipment that no longer fits your system and replace it with a more appropriate implement. You may wind up selling obsolete equipment for less than you paid for it, but that's not nearly as great a price as you would pay to continue operating inefficiently.

An example of this would be a grower who purchases a Cub tractor at a good price when he is first starting out. This tractor is good at cultivating one row, and fits well in a system where wide rows and push planters are used. But then the grower decides to expand and move to a more efficient narrow-row system that requires a multi-row cultivator. The Cub, good as it is, suddenly becomes obsolete. It was designed and built for one-row cultivation and is not wide enough for proper multi-row cultivation.

I believe all equipment, including tractors, should be viewed as tools designed by engineers to perform certain specific jobs well. Before buying any piece of equipment, you should first decide what job needs to be performed. Then go out and find the proper tool for that job.

To do an efficient job of growing 10 acres of vegetables, more or less, you will need three tractors. The main features and functions of these tractors are:

EQUIPPING YOUR WHATLEY-STYLE FARM, SENSIBLY AND ECONOMICALLY

Whether you go to an auction, an equipment dealer or buy directly from another farmer, the "used" market is a good source of inexpensive tractors and other farm equipment.

Big Tractor — 40- to 50-hp, live hydraulics:

- Primary tillage
- Manure handling
- Help with secondary tillage
- Operate irrigation pump

Second Tractor — 30- to 45-hp, 3-point hitch, wide front end:

- Planting
- Operate specialty implements
- Cultivation (of 30-inch rows)
- Secondary tillage

Cultivating Tractor — 18- to 30-hp, wide front end, mid-mount cultivator capacity:

- Cultivation of 15-inch rows and between beds

Buying three tractors of these types new would cost about $40,000. That is way too much for most growers. Buy used equipment and the cost will be only about $10,000.

Big Tractor

Since your largest tractor will be doing all of the primary tillage, it must be able to pull at least a 2-bottom plow, and perferably a 3-bottom plow. That requires 40 to 50 hp. If manure is to be spread on the fields to maintain their tilth and increase fertility, a front end loader is a must. To greatly help speed manure-handling chores, you will want to have live hydraulics, which allow you to operate the loader independently of the clutch. Power steering is not an absolute necessity, but it will make loader work easier. A wide front end and a 3-point hitch aren't absolutely necessary, but they sure would come in handy in assisting the second tractor during peak work periods in spring.

Your big tractor will cost from $2,000 to $5,000. The cheaper ones will usually be without both the 3-point hitch and a wide front end. If you take your time, you can oftentimes find one with those two items in the $3,000 to $4,000 range.

The following is a listing of some of the tractors from each major manufacturer that fulfill the requirements for your big tractor. It also includes price ranges, horsepower ratings, model years and price ranges as listed in the 1987 edition of the "Quick Reference Guide For Farm Tractors And Combines," which is published by Hot Line Inc., P.O. Box 1052, Fort Dodge, Iowa, 50501. Information on types and availability of hitches and live hydraulics comes from many sources — from personal experience to tractor salesmen — and is not mistake-proof.

Big Tractors

MODEL	YEARS MADE	HP	HITCH TYPE	LIVE HYDRAULICS	FRONT END	PRICE RANGE (1987)
Allis Chalmers						
WD	1948-53	40	Snap coupler	No	Most narrow	$ 100- 925
WD45	1953-57	45	Snap coupler	No	Most narrow	325-1,250
D-15	1960-67	40	Later models have 3-point	Yes	Both	2,050-2,725
D-17	1958-67	54	Later models have 3-point		Both	1,075-3,200
160	1970-74	40	3-point	Yes	Most wide	2,000-3,700
170	1968-73	54	3-point	Yes	Most wide	1,620-5,380
John Deere						
A	1934-52	39		No	Most narrow	250- 800
60	1952-56	42	No draft 3-point control	Yes	Most narrow	625- 900
520	1956-58	39	3-point	Yes	Most	600-1,500
620	1956-58	49	3-point	Yes	Narrow	800-1,600
530	1958-66	41	3-point	Yes	Most narrow	1,000-1,800
2010	1960-65	47	3-point	Yes	Both	400-2,100
2020	1965-71	54	3-point	Yes	Most wide	1,000-4,100
2510	1966-68	54	3-point	Yes	Most wide	2,200-3,975
1520	1968-73	48	3-point	Yes	Most wide	1,000-3,920
Ford						
841	1957-61	45	3-point	No	Wide	880-2,280
941	1957-61	47	3-point	No	Wide	900-1,000
3600	1975-81	41	3-point	Yes	Wide	2,900-5,000
4000	1963-67	52	3-point	Yes	Wide	1,000-5,600
5000	1965-82	48	3-point	Yes	Wide	1,200-6,500
International						
H	1939-53	35	Fast hitch	Some late models	Most narrow	400-1,500
M	1939-53	49	Fast hitch	Some late models	Most narrow	600-2,000
F-300	1954-56	40	2-point	Yes	Both	320-1,200
F-350	1956-58	42	2-point	Yes	Both	320-1,960
F-460	1958-63	52	2-point	Yes	Both	320-1,600
F-504	1961-68	46	3-point	Yes	Both	750-2,350
F-544	1968-73	54	3-point	Yes	Both	1,600-4,000
Massey Ferguson						
65	1958-64	48	3-point	Yes	Wide	900-2,400
165	1964-75	52	3-point	Yes	Wide	1,600-3,650
J.I. Case						
400	1955-58	51	Eagle	Yes	Both	550-1,200
500B	1958-59	47	Eagle	Yes	Both	350-1,000
530	1960-69	41	Later models have 3-point	Yes	Both	650-3,600

Second Tractor

You will quickly see that your second tractor will log the most hours of use. To perform its many duties well, it will need to have just the proper options. You will need 30 to 45 horsepower to pull secondary tillage implements. A 3-point hitch is mandatory so that you can operate specialty implements such as bedmakers and mulch layers. And, it must be able to straddle beds. That calls for a wide front end.

The 3-point hitch is the limiting factor when looking for a tractor in this horsepower range. Automaker Henry Ford and British inventor Harry Ferguson first introduced the 3-point hitch with the "Ferguson System" of linkage and draft control on the Ford 9N tractor in 1939. Every Ford and Ferguson farm tractor made since then came with a 3-point hitch. While other tractor companies had their own systems of attaching implements, the 3-point hitch was, by far, the best. By the mid-'50s and early '60s, the other tractormakers began adopting the 3-point hitch.

Here are the specifications on popular tractors that meet the requirements for your second tractor:

Second Tractors

MODEL	YEARS MADE	HP	HITCH TYPE	LIVE HYDRAULICS	FRONT END	PRICE RANGE (1987)
Allis Chalmers						
160	1970-74	40	3-point	Yes	Most Wide	$2,000-3,700
5040	1976-present	40	3-point	Yes	Most Wide	3,500-7,000
J.I. Case						
430	1960-1969	34	Later models have 3-point	Yes	Most Wide	720-1,750
John Deere						
520	1956-58	39	3-point	Yes	Both	600-1,500
820	1956-58	34	3-point	Yes	Both	900-2,500
1010	1960-65	36	3-point	Yes	Both	600-1,900
1020	1965-73	39	3-point	Yes	Both	1,250-3,470
2040	1976-82	41	3-point	Yes	Both	4,500-9,500
Ford						
650	1954-57	35	3-point	Yes	Wide	450-2,000
641	1957-61	34	3-point	Yes	Wide	1,100-3,250
2100	1967-75	32	3-point	Yes	Wide	760-3,700
3100	1966-75	38	3-point	Yes	Wide	960-3,550
International						
404	1961-68	37	3-point	Yes	Both	600-2,100
424	1964-67	37	3-point	Yes	Both	960-2,250
444	1967-71	38	3-point	Yes	Both	800-2,950
454	1970-73	41	3-point	Yes	Both	1,200-2,250
484	1978-84	41	3-point	Yes	Both	3,600-7,200
Massey Ferguson						
MF50	1958-64	33	3-point	Yes	Wide	800-1,300
TO-35	1958-60	35	3-point	Yes	Wide	1,000-2,500
MF35	1958-64	33	3-point	Yes	Wide	850-1,775
MF135	1964-75	35	3-point	Yes	Wide	1,800-3,000
MF150	1964-75	38	3-point	Yes	Wide	1,050-2,700
MF235	1975-76	41	3-point	Yes	Wide	1,775-4,350
MF245	1976-83	43	3-point	Yes	Wide	2,425-8,050

Cultivating Tractor

Your cultivating tractor must have a wide front end and mid-mount cultivators to work right next to the crops. Mid-mount cultivators are better than rear-mounted cultivators because they are more stable, with virtually no side sway, and provide better depth control. They also allow tractor drivers to see exactly where the cultivators are running, without turning around.

2-Tractor System

For starting up a new operation or farming less than 10 acres of vegetables, two tractors can be used, instead of three. The second tractor and the cultivating tractor would be the ones to buy. The second tractor becomes the workhorse and takes on the duties of the big tractor. Your plow must fit the second tractor. Make sure this tractor also has live hydraulics to operate a front end loader.

However, if you are starting up a new operation and plan to expand quickly and buy your big tractor in a few years, then you could do without the loader for this mid-sized tractor. You could farm the first few years without spreading manure or possibly pay a neighbor to spread it for you.

You can and may want to try to farm 10 or more acres with just these two tractors, but I believe it will be more expensive because of missed planting dates and higher weeding costs. In spring, the workload on a diversified vegetable farm is quite heavy. Crops have to be planted every week. Rains often keep the soil unfit for any type of tractor work for a week or longer. This means that you are often forced to try to do as much tractor work as possible in a very short period of time. If you are trying to do all of your soil preparation and planting with just the mid-sized tractor, you will not have enough time to do the job properly. This usually results in missed plantings that later cost you money in lost sales, or in poor soil preparation that later causes severe weed control problems.

Cultivating Tractors

MODEL	YEARS MADE	HP	ROWS CULTIVATED	TRACTOR CONFIGURATION	FRONT END	PRICE RANGE (1987)
Allis Chalmers						
B	1938-57	23	1	Std.	Both	$ 285- 650
G	1948-55	10	1	Rear Engine	Wide	300-1,050
D-10	1959-67	29	2	Std.	Both	450-1,050
D-12	1959-67	29	2	Std.	Both	560- 950
J.I. Case						
VA	1953-55	22	2	Std.	Both	400-1,000
VAC	1953-?	22	2	Std.	Both	160- 600
200B	1956-57	31	2	Std.	Both	250- 625
John Deere						
B	1940-52	29	2	Std.	Both	150- 950
L	1940-46	9	1	Offset	Wide	100- 400
LA	1941-46	13	1	Offset	Wide	300- 600
40	1953-55	25	2	Std.	Both	750- 900
420	1956-58	29	2	Std.	Both	550-1,475
430	1958-60	30	2	Std.	Both	800-1,500

Ford
No tractors with mid-mount cultivators, except one offset tractor made in 1958, but it is rare and parts are unavailable.

MODEL	YEARS MADE	HP	ROWS CULTIVATED	TRACTOR CONFIGURATION	FRONT END	PRICE RANGE (1987)
International						
A, AV	1939-47	20	1	Offset	Wide	400-1,200
Super A	1947-54	21	1	Offset	Wide	1,150-2,025
C	1948-51	22	2	Std.	Most Narrow	350-1,200
Super C	1951-54	24	2	Std.	Most Narrow	375-1,250
F100	1954-56	21	1	Offset	Wide	425- 525
F130	1956-58	23	1	Offset	Wide	450- 550
F140	1958-79	24	1	Offset	Wide	2,220-3,000

Massey Ferguson
Made 2- and 4-row mid-mount cultivators for their larger tractors the 50, 150, 65 and 165. None were made for smaller tractors.

EQUIPPING YOUR WHATLEY-STYLE FARM, SENSIBLY AND ECONOMICALLY

1-Tractor System

For farming part-time on five acres or less, or for someone who wants to start out slowly, just one tractor can be used. This tractor will have to be chosen very carefully because it must be able to perform all the jobs that three tractors would do on a larger farm. Such a tractor must have:

- 3-point hitch to operate specialty vegetable implements such as bedmakers and plastic layers.
- Wide front end to straddle beds.
- Mid-mount cultivators for proper cultivation.

That quickly narrows your choices to just a few tractors:

- Allis Chalmers D-12 (later models)
- John Deere 40, 420, 430, 1010
- Case 430, VAC (later models)
- Massey Ferguson 50, 150

Probably a few more tractors would meet the requirements for a one-tractor operation. The problem is finding them with a set of cultivators. Your chances of finding a tractor equipped with cultivators are much better with the smaller tractors like the JD 420, Case VAC and AC D-12, which were often originally purchased as cultivating tractors. One of the smaller tractors then could become your cultivating tractor when you expand your operation. If you bought a MF 50 or 150 with cultivators, it could later become your second tractor or remain a cultivating machine. Just plan ahead and try to look at your farm, both now and in the future, as an entire system.

Implements

The implements needed for a vegetable operation can be divided into five main categories:

1. Primary tillage
2. Secondary tillage
3. Planters
4. Cultivators
5. Specialty implements
   Bedmakers
   Plastic layers
   Transplanters

Primary Tillage

There are three types of primary tillage— *moldboard plowing, rotovation and chisel plowing*. Moldboard plowing and rotovation are the main methods used on vegetable operations because they cover trash well and provide a good surface in which to prepare the fine seedbed necessary for the proper planting of vegetable seed. Chisel plowing, on the other hand, does not cover trash well and is not used on most vegetable farms. Perhaps an even greater drawback of chisel plowing is the fact that it requires a lot of horsepower. Each chisel shank translates into about 15 horsepower, which means even a small chisel plow with five shanks would need a 75-hp tractor.

The moldboard plow has earned a nasty reputation as the cause of serious soil erosion. When properly used, though, it is just the thing for plowing down soil-improving green manures like the stand of red clover in this photo.

Moldboard plowing is a more common method of primary tillage than rotovation bacause it is faster, the equipment is cheaper to purchase and operate, and it causes less soil crusting than rotovation. A used 2-bottom or 3-bottom 16-inch moldboard plow, for example, costs $200 to $600. A rotovator, which most likely would have to be purchased new, costs $1,200 to $1,500.

Secondary Tillage

The purpose of secondary tillage is to break down large clods left by primary tillage and to pack down the soil to produce a firm seedbed. Secondary tillage implements include the following:

- *Single-action disk-harrow*— does a poor job of soil preparation. I don't recommend buying one.

- *Double-action disk-harrow (or tandem disk)*— is the most commonly used secondary tillage implement. It does a good job of breaking down clods and forming the seedbed. Buy one eight to nine feet wide. They are available as pull-types with transport wheels. New, they are very expensive, but the supply of used tandem disks is usually good.

A pull-type, eight-foot tandem disk costs $200 to $500, used. A used transport disk of this type, complete with hydraulic cylinder and hoses, costs $400 to $900.

• *Offset disk-harrow*— is not used much on small vegetable farms. It covers trash better than a tandem disk, but also requires more horsepower. A heavy offset disk requires nine to 10 horsepower per foot of disk, while a heavy tandem disk requires only five to six horsepower per foot. This means that a farmer who has a 50-hp big tractor would be limited to a five-foot offset disk, which is a difficult item to find. Buy a tandem disk.

• *Spike tooth harrow*— is usually used for leveling the soil after disking. Use is limited. Don't buy one.

• *Spring tooth harrow*— is also used for leveling the soil after disking. It is also quite good for freshening a stale seedbed before planting. It digs deeper than a spike tooth harrow and does a better job of removing weeds. Established growers often use these on their farms, but it is not a "must have" item for starting up a new operation. A used nine- to 12-foot harrow costs $300 to $400 for a pull-type and $400 to $900 for a 3-point hitch type.

• *Packer Wheels*— are usually pulled behind a tandem disk to firm and level the seedbed. They should be purchased one foot longer than your tandem disk. This is a "must have" item. Buy one.

A used nine- to 11-foot packer wheel costs $300 to $800. Of course, you can get by much more economically and still achieve good results by simply using large bolts, washers and chain to pull a log or section of telephone pole behind your disk.

• *Cultimulchers*— are specialty implements that combine a spring tooth harrow and a packer wheel. They do a good job of leveling and firming the seedbed. Not necessary for a start-up operation. Cost of a used nine-foot cultimulcher is $800 to $2,000.

The key to getting the most out of manure is to incorporate it as quickly as possible after spreading.

Planters

Push planters are, as the name implies, pushed by hand. They are inexpensive and will plant seed adequately on up to half an acre. On a larger scale, tractor-drawn planters are preferred because push planters cannot make rows straight enough for effective mechanized cultivation. Push planters range in price from about $75 for a Precision-brand planter to $350 for a Planet Jr. planter.

Tractor-drawn planters come in seven types:

1. Pneumatic
2. Plate
3. Plateless
4. Fluid
5. Precision
6. Belt
7. Planet Jr.

The new plateless and pneumatic planters are a bit more accurate than the old plate-type planters, but they are also expensive. These planters are so new that few are available used. That is why I recommend buying an older plate-type planter. This is primarily a large seed planter and will be used to plant sweet corn. Buy a 2- or 4-row unit. A used 2-row planter in good condition is hard to find. Four-row units are much more common. Buy a name brand like International, John Deere or Allis Chalmers, but check with a dealer about parts availability, first. Cost is $200 to $600.

Fluid planters plant pregerminated seed in a band of gelatin. They are still experimental, but look promising.

Precision plate, belt and Planet Jr. planters were all developed primarily to plant small-seeded crops such as carrots and lettuce. The precision plate planter and the belt planters both allow you to plant a single small seed at precisely the spacing you want. Precision planters do this by using finely cut, stainless steel seed plates to meter out seed, while the belt planters use belts punched with holes. Plate or belt speed is changed by the various gears. Both types of planter are expensive. Precision plate planters also require a heavy investment in seed plates, which cost $30 to $60 each.

The Planet Jr. planter augers seed out through a choice of 39 holes of different sizes in metal plates. The gearing remains the same, so the only variable is the size of the hole. This does not allow precision spacing of seed. Instead, it is a coarse metering system with the ability to plant a solid row of seed, thinly or thickly. These planters are inexpensive and seed settings are easier to change than on precision and belt planters.

The precision and belt planters were primarily developed to reduce the need for labor in thinning crops. They also save money by reducing seed costs and increasing yield by minimizing the number of culls caused by overly close spacing. The savings on thinning assumes the use of herbicides to control weeds. If herbicides are not used, the labor required for weed control will negate any cost savings from the use of these planters. When planting without herbicides, it is cheaper and better to solid-seed the crop. There are three reasons for that:

1. Solid-seeded crops compete better against weeds than plants spaced 12 inches apart.

2. Cultivation is faster and easier. Because of the higher plant populations in solid-seeded rows, knocking out some crop plants with the cultivator is not a big concern. Cultivators can be run much closer to crop plants and at higher speeds. In precision-spaced plantings, cultivation is often tedious because you have to worry about removing crop plants and leaving gaps in the row.

3. Hand-hoeing and thinning a solid-planted row is faster and easier than trying to clean up a weed-infested row with crop plants a foot apart. With solid-seeded rows, no time is wasted trying to distinguish crop plants from weeds.

For those reasons, I recommend buying a Planet Jr. planter to plant small-seeded crops. A 3-row unit on a toolbar set up for a 3-point hitch costs about $1,200 new. It will plant both wide and narrow rows. The middle unit is left empty when planting the wide rows. This includes the brushes necessary for small seeds and a 2-inch scatter shoe for proper spacing of root crops such as carrots, beets and radishes.

John Deere 33

This precision plate planter costs about $3,000 for a basic 3-row unit on a 3-point hitch. Then you will have to buy seed plates. For a diversified farm having many crops, these plates will cost $800 to $1,000.

Stanhay Belt Planter

Complete with all the options, this planter will cost about $4,200 for a 3-row unit. This includes a cost of $1,450 for land wheels, which are necessary when planting on beds. Land wheels are not necessary if you're not using a bed system.

Cultivation And Cultivators

Vegetable crops are fairly specific in the herbicides that can be used on them. That's why planting many different crops requires many different herbicides. A number of relatively small patches of different crops planted at various dates can drive you crazy trying to match them with the proper herbicide. This is expensive in both time and money. Many vegetable crops are also very sensitive to herbicide carryover, and using a lot of herbicides greatly limits your ability to rotate crops. Rotating vegetable crops is very important for disease, insect and weed control.

Without herbicides, the main defenses against weeds will be black plastic mulch, hand hoeing and mechanical cultivation. Black plastic is used on beds of warm-season crops such as tomatoes, peppers, eggplant and all of the cucurbits. Hand-hoeing, which is self-explanatory, is a last resort. Mechanical cultivation thus becomes your main defense against weeds.

Although handy, rear-mounted cultivators do not provide the stability and precision of mid-mounted rigs.

To properly cultivate, every field operation prior to cultivation must be performed properly. In order to accomplish this, a grower must have the right machinery to ensure sufficient time to perform all the field operations during the peak work period in spring. This peak is so intense that some growers use as many as five tractors on a 30-acre farm.

The conditions necessary for good cultivation are:

1. Optimum fertility for fast-growing crop plants.
2. Fine, smooth and level seedbed.
3. Good, uniform crop stand at the proper stage of growth.
4. Straight rows with accurate spacing.
5. Optimum soil moisture.
6. Enough time to cultivate when conditions are right, which might be only for a couple of hours in one afternoon.
7. Using the proper cultivators for each crop at the correct speed, depth and spacing.

Mid-Mounted Cultivators

Mounted under the belly of the tractor, as the name implies, mid-mount cultivators are raised by hydraulics or mechanically with a large lever. The cultivating tractor should be purchased with a set of cultivators. When fitted with sweeps, these cultivators will be used to cultivate between beds of black plastic. Sweeps usually work better than shovels in throwing soil onto the buried edge of black plastic to kill weeds growing there.

The cultivating tractor, when fitted with a specialty mid-mount cultivator, will also cultivate the narrow rows of direct-seeded and transplanted crops. Many custom welding shops make specialty cultivators for narrow rows. These units are quickly and easily attached to tractors. Most of these shops are located in the vegetable growing regions of California, Florida, Michigan and New Jersey. They are most commonly located by word of mouth, or sometimes by ads in local newspapers. Most of the cultivators made by these firms are smaller and can be arranged in many different ways to suit specific soil and crop needs.

Custom-built cultivators will vary widely in price, depending on their complexity of construction. They cost from $500 to $1,000. One unique type of cultivator which I have used with good success is the rolling cage cultivator made by the Buddingh Weeder Co. This cultivator does an excellent job of removing weeds in the two- to four-leaf stage. It is inexpensive and easy to mount and dismount. A Buddingh wheel hoe costs about $500 and is available from:

Buddingh Weeder Co.
7015 Hammond Ave.
Dutton, Mich. 49316
(616) 698-8316

Rear-Mounted Cultivators

Rear-mounted cultivators are attached to the tractor by the 3-point hitch. They are not as precise as the mid-mounted cultivators because side sway cannot be completely eliminated from rear-mounted implements. Sway blocks or bars on the tractor and stabilizing disks or shoes on the cultivator keep the rear-mounted cultivator steady enough to do a good job on wide-row, large-seeded crops such as corn, and transplanted crops on wide rows.

When buying used cultivators, which I recommend because they are readily available at half the price of new cultivators, make sure that the one you buy has shields with it. Shields alone are not usually available in the used market. Buying new shields would cost almost as much as a new cultivator. Another useful attachment is a set of 14-inch disks-on-shanks that will fit your cultivator. These disks do a good job of hilling sweet corn on last cultivation. Cost of a used, 2-row cultivator with shields should be $200 to $250.

Rotary Hoe

This implement is an absolute necessity for controlling weeds with only a modicum of herbicides. It breaks up crusted soil on large-seeded crops such as corn, peas and beans, and controls weeds in the row between young crop plants.

The rotary hoe is used before the crop is large enough for regular cultivation. It works by kicking out the small-seeded weeds as they germinate, leaving unharmed the large-seeded crop plants. Rotary hoes work best when the soil is crusted and weeds are just emerging. After weeds have reached the two-leaf stage, the rotary hoe cannot remove them, so timeliness is very important.

There are two types of rotary hoes. The newer, spring-loaded type is mounted on a toolbar with a 3-point hitch attachment. The older, pull-type hoe is not spring-loaded. The spring-loaded hoe is much superior because the

Notched, rolling shields protect young crop plants during cultivation, but also destroy weeds right next to the row. Cultivator shields also come in rigid types, with open or closed tops.

The hundreds of spoons on a rotary hoe quickly pulverize soil crust and uproot small weeds. If you're farming with few herbicides, a rotary hoe will be one of your best machinery buys.

springs allow each hoe on the implement to float independently, providing better coverage on uneven ground. And, with 3-point hitch mounting, you can easily control the depth of soil penetration from the tractor seat. This optimizes the effectiveness of the implement by allowing the operator to quickly adjust the depth for varying soil conditions.

In contrast, controlling the depth of penetration on the old, pull-type hoe is much more difficult, since it involves adding weights to the implement's frame. The other weakness of the pull-type hoe is that it does not float on uneven ground. But, despite all that, the older rotary hoe still does an adequate job. When adapted to a 3-point hitch, it could become one of your best machinery buys. The spring-loaded hoe is still so new that used prices are high and supply is scarce. A 15-foot spring-loaded hoe (the most common size) in good condition costs $1,200 to $1,500. A pull-type hoe of nine to 12 feet in width costs $200 to $500.

Transplanters

Mulch transplanters— are the most versatile of transplanters. They can transplant crops through black plastic and on bare ground. And one company, the Mechanical Transplanter Co., has an attachment for their Model 900 machine that allows you to plant seeds, too. A one-row machine could then transplant tomatoes, eggplant and the early vine crops and also be used to seed the main season vine crops. This leaves only cole crops, lettuce and peppers to be transplanted by hand. The cost of a one-row machine from Mechanical Transplanter is about $2,000, complete with the Jiffy Plug mix attachment for seeds. I have never used one of these machines, but have only talked to a grower who owns one. If the machine can transplant and plant seed as well as reported, I think it would be worth the money. Transplanter makers include:

Mechanical Transplanter Co.
Box 10088
1150 S. Central Ave.
Holland, Mich. 49423
(616) 396-8738

Holland Transplanter Co.
510 E. 16th St.
Box 535
Holland, Mich. 49423
(616) 392-3579

Finger-type transplanters— have a set of mechanical fingers into which transplants are placed. These fingers do the planting. There are two types, those designed primarily for stem plants such as tomatoes and peppers, and those designed primarily for leafy crops such as lettuce and celery. The major drawback of these planters is that they cannot be used with black plastic. This effectively limits their use to transplanting the cole crops and lettuce. Unless a large acreage of these crops is to be planted, it is not worthwhile to buy a finger-type planter. This machine is also available from Mechanical Transplanter Co. and Holland Transplanter Co.

Old-style transplanters— are the type with which you actually set the plant in the ground by hand, right behind the furrow opener. The advantage of this is that you can plant any size plant at any spacing in the row, without having to adjust the machine. With the mulch and finger-type planters, it takes time to readjust the machine every time you want to change the plant spacing. Use of an old-style planter, like the finger-type planter, is limited because it cannot plant through black plastic. A one-row unit costs about $700. They are made by Holland Transplanter Co. and:

Ellis Mfg. Co. Inc.
107 W. Railroad St.
Box 246
Verona, Wis. 53593
(608) 845-6472

Front-end loader— is necessary for handling manure and other bulky materials. A new loader costs about $2,800, while a used loader costs about $1,500. Used loaders are often hard to find. Start your search for one with a machinery dealer who may have one sitting out back.

Manure spreaders— A good, used manure spreader, like a good used loader, is an item in great demand and may have to be purchased at an auction. Figure on spending about $1,000. Buying a name brand popular in your area will make it easier to find parts.

Sprayers—Depending on your choice of pest and disease control methods, your sprayer needs could range from a $120 backpack sprayer to a pair of tractor-drawn 50- to 100-gallon sprayers costing $2,000.

Basic Start-Up Machinery

The following is a list of the basic, start-up equipment needed to efficiently manage a 10-acre diversified vegetable operation. Prices are estimates for good, used equipment.

Second tractor (30- to 45-hp)	$ 4,500
Cultivating tractor	1,500
Plow	500
Disk	800
Packer wheel	250
Rotary hoe	800
Specialty cultivator for narrow rows	500
Rear-mounted cultivator for wide rows	250
Bedmaker	600
Plastic mulch layer	1,000
Planter (plate-type) for sweet corn	500
Planter (Planet Jr.) for other seeds	1,200
Sprayer	1,000
	Total $13,400

To increase the acreage planted to 20, you would need a 40- to 50-hp tractor costing about $4,500, plus a manure spreader ($1,000) and front end loader ($1,500) for a grand total of about $20,400.

Tom Morris, a former small-scale vegetable grower and farm manager of the 305-acre Rodale Research Center in Maxatawny, Pa., is completing his master's degree in soil fertility at the University of Connecticut.

Bid Better, Buy Cheaper

A veteran auction-goer explains how to leave a sale with the machine you want and money in your pocket.

JOHN G. RUFF

You don't need someone to tell you that farm machinery costs a lot more today than it did a decade ago. What may surprise you is how much more it costs. Since 1977, machinery prices have increased more than any other variable costs (including fertilizers and pesticides): nearly 80 percent, according to USDA.

During the same time, auction prices for used equipment have fallen drastically, often to nearly half what you'd pay for the same used item from a dealer. I recently saw a clean IHC 856 tractor, with cab and air conditioning, sell for less than $3,000 at an auction. The price would have easily been $7,000 or $8,000 a few years ago.

As a result, even farmers who once avoided auctions now see them as an excellent way to stretch their machinery budgets, or to expand and upgrade their equipment inventories.

Change With The Times

Are auctions the cheapest places to find good equipment? They can be, but only if you adapt your bidding and buying habits to today's conditions.

My father and I have bought most of the machinery for our farms at auctions during the past 35 years. With careful planning and a little willpower, we've saved hundreds of dollars on items that have given us many years of use. A 10-foot Wilbek offset disk I bought in 1976 is just one example. I paid just $650 for it; similar ones were going for $1,000-plus at other auctions. Yet it needed no repairs at all for three years.

In many ways, auctions haven't changed much. You can still buy at a lower price than you'd pay a dealer. You still have to pay cash. And you don't get a guarantee with your merchandise.

But one important aspect has changed: the reliability of the equipment you buy at auctions. Not long ago, you could buy an item and, with reasonable care, use it for several seasons before major repairs would be needed. Those days are gone. Today, we buy a machine, try to figure out what's wrong with it, and hope we can make most of the repairs ourselves without buying a lot of new parts.

The cause of this change is easy to pinpoint. More and more farmers are facing serious cash-flow problems, and often are forced to neglect routine maintenance. In some cases, they don't even have the money to make necessary repairs.

EQUIPPING YOUR WHATLEY-STYLE FARM, SENSIBLY AND ECONOMICALLY

You may not be able to entirely avoid buying someone else's machinery headache. But you can minimize your risk by considering your auction visit a business trip instead of a Saturday outing.

To begin with, you need to determine just how vital a particular piece of machinery is to your operation, and whether or not an auction is the best place to find it. We've found it useful to group equipment purchases into three classifications.

First is the machine you need desperately, but aren't likely to find being used by other local farmers. Waiting for one to show up on a sale bill could end up costing you money instead of saving it. Better try a dealer.

Second is the machine you can postpone buying until you find the right kind at the right price. Auctions can be a very good source for these types of purchases.

Finally, there's the unit about which you might say: "I sure wish I had one, but it's really got to sell cheap." If you attend a lot of auctions, you'll be surprised at how many of these wish-list items you can acquire over a period of several years.

It's also good to set a limit on how far you'll drive to an auction, and to subscribe to a regional paper that carries sale bills for that area. Since I make as many as three trips to the sale site (I'll explain why in a minute), I generally won't attend an auction more than 50 or 100 miles from my farm. But to be sure I don't miss anything important, I subscribe to four newspapers (two local and two regional), and carefully read about 1,500 sale bills a year.

Try Before You Buy

When you spot an attractive sale bill, determine whether the types and brands of machinery to be sold will fit your needs, and familiarize yourself with the used-market prices for these items.

Are parts available from a local dealer, or will you have to stock them yourself by ordering from the manufacturer? Does it look like the manufacturer, or the dealer who carries the parts, will still be in business several years from now? You must have answers to these questions before preparing to visit the auction.

Suppose you read a sale bill listing a machine you want. You've determined what the item is worth, and you feel you have money to buy it. An established local dealer can supply parts, and you have a good idea of the type and cost of repairs commonly needed for the unit.

The next step is deciding whether you actually want to bid on the machine. If possible, visit the sale site beforehand and inspect the unit. Better yet, try it out. Contact the dealer who has sold parts or made repairs on it. You may find out it's not such a bargain, after all.

If the machine appears field-worthy, it's time to calculate how much you're willing to pay for it. Years ago, we used to base our highest bid on the used-market price, with a small allowance for repairs. Today, we assume from the start that repairs will be needed.

I can't overemphasize how important this is, given how quickly parts and repair costs are increasing. It seems manufacturers realize that the only way to sell new machinery is to make the older models too costly to fix.

After our pre-sale inspection, we're usually able to estimate what the repairs will be and how much they'll cost. We subtract that amount from our estimate of the machine's actual value. The difference is what we can afford to bid.

Here again, there will be times when auction machinery will turn out to be more expensive than you think. For example, suppose you buy a tractor for $4,000 less than a dealer would ask. It smokes a bit, and the tires are worn, but you couldn't resist the price. After you get it home, though, you might notice the engine and transmission need complete overhauls. Together, those repairs could easily total $4,000 — before you even think about tires, batteries and other incidentals.

Spot Warning Signs Early

The last, and possibly most difficult, step is the auction, itself. Go early enough to inspect most of the machinery being sold. If much of it shows signs of misuse and neglected repairs, the unit you want may also have been mistreated. If it appears in good shape, you may want to raise your top bid for the machine. (Remember, too, that even new machinery can be quickly ruined by a careless owner.)

Try to talk with the seller's neighbors. They can often provide helpful insights. And during the sale, pay close attention to the bidding on other items. A lot of neighbors bidding on machinery is a good sign; very few people bidding is a bad one.

Of course, the most important step is your bid. It's really very easy to buy a machine at an auction. It just costs money. Keep raising your bid, and you'll eventually own anything you want.

But you don't save money at an auction by buying a lot. You save it by buying carefully. That means following some rules. One of our favorites is: Never continue bidding unless there's someone else competing with you. The old saying is that it only takes two bidders to make an auction. That's wrong. It only takes one bidder and a clever auctioneer to run the bid way up.

An auctioneer friend of mine once pointed out a few of the tricks of his trade during a sale at which he was off duty:

• If the auctioneer says he has an opening bid but can't get a raise — then decides not to sell the item — chances are he never had the opening bid in the first place. Try offering a lower bid; you might find it's accepted.

• If you can't see who is bidding against you, ask to see the other bidder (standing in the back can improve your view of the crowd). Sometimes it's the auctioneer, himself, bidding for an absentee bidder. A reputable auctioneer will tell you this.

• If an auctioneer occasionally backs up a bid (for example, he drops to $80 after claiming he had $100), it could be just an honest mistake. If he does it a lot, he probably got caught trying to run the bid. Likewise, a fast bid return may mean a bid is being run. Look for the other bidder if this continues for a long time.

There's a chance the owner arranged for someone to run the bid up to a minimum reserve price. I get suspicious if the owner or his family or neighbors bid against me, particularly if their bids seem far higher than an item is worth. Don't get caught in this game unless you're willing to pay the higher price.

On the other hand, if your competition is a dealer, his bid will be a good indication of the item's true worth, because it's based on his expected profit margin plus repairs. Even if you have to exceed your maximum bid to buy the item, chances are you'll still get a competitive price.

Another important rule is to always stop bidding when you reach your predetermined limit. This can be difficult if you become affected by "auction fever." Competent auctioneers can make it nearly impossible to say no to the "buy of a lifetime," because it's so easy to "bid just one more time."

You must learn to recognize those and similar phrases for what they are: sales pitches. Remember that the auctioneer is working for the owner. His job is to obtain the most money for each item on sale, not provide you with an easy source of low-cost machinery.

Don't expect him to work for you. Set your top limit, and stick to it. There will almost always be another opportunity to buy the item — and you'll probably end up with a better price.

Nobody's Perfect!

Does this system work? We've used it for many years, and have been well-satisfied with most of our purchases. Repair costs have been in line with what we expected for normal wear and tear, and virtually everything we've bought has paid for itself in savings within just a few years.

Of course, there are exceptions. One of my favorite, and ultimately most useful, auction purchases was an old Dearborn self-propelled swather I bought on impulse for $500 in 1972. The price was about one-fourth what I would have paid for it at a dealer. But I soon found out that Dearborn, the company which originally sold the machine, had dropped its swather line. I ended up with a set of parts books from the manufacturer, but I had to determine for myself which of those parts fit my unit.

The price included the frame and a lot of worn-out parts, so we at least knew where to install the repair parts. I was able to pick up more than $1,000 worth of parts for $100 at an auction by a Dearborn dealer who was going out of business. But in all, I've probably spent more than $4,000 (including an engine replacement) getting the unit in shape.

Much of that cost came during the first few years, during which a neighbor shared the use and expenses with me. But I still have to stock my own parts inventory to avoid downtime during haymaking.

Despite all this, it has been an excellent and comparatively cheap swather to own and operate. My repair costs average about $1 to $1.50 per acre, based on the 250 acres of hay I cut yearly.

Fortunately, I've only made this type of mistake once. I hope you can avoid it entirely.

John Ruff raises beef cattle and hogs on his 670-acre farm near Logan, Kan.

How To Buy A *Good* Used Tractor

S. E. DURCHOLZ

Buying a used tractor may seem riskier than buying a used car. But by being a patient, careful shopper, you can put a highly serviceable tractor in the machine shed — without bringing another farmer's troubles home.

"There is a lot of good equipment out there — if you look for it," says Sam Huber, an Extension agricultural engineer at Ohio State University. Considering several tractor brands will improve your choices.

The best place to start looking for a *good* used tractor may be right in your own hometown. "A reputable dealer is about as good a place as any," says Leonard Bashford, a professor of agricultural engineering at the University of Nebraska in Lincoln. A farm sale is OK, provided you've done the necessary homework on the tractor before it goes on the auction block.

The best way I've found to track down a *good* used tractor is to keep a large notebook (shirt pocket notebooks are just too small) containing all the vital statistics on the tractors under consideration. Then, when it comes time for the final decision, you have the good and bad points of each tractor at your fingertips.

Whether you're buying a trade-in from a dealer or directly from another farmer, personal experience and the experts agree on one thing — get to know the previous owner.

"It's always nice to know why someone traded it off. Usually it's because they wanted a larger tractor or the one they had couldn't do something they wanted it to," says OSU's Huber.

Keep your eyes open when you go out to another farm to see the tractor advertised in the Sunday paper, Bashford adds. "Look around . . . If he doesn't take care of his other machines, he probably doesn't take care of that tractor. Talk to the farmer about what he's done to it. It's like buying a used car. Learn what's likely to go wrong soon," he says.

A newer tractor is easier to evaluate. A tractor 10 to 15 years old is just like an older car, Huber says. "It depends on who had it. Could be someone took care of it and even with 100,000 miles it's still OK. Or it could possibly be a pile of junk."

It's hard to tell the exact condition of a tractor without tearing it completely apart, but there are many exterior clues and some simple mechanical tests that will keep you on the right track.

"Check the oil, you're darn right. Pull the dipstick out. If there's no oil in there or it's old and dirty, it makes you kind of wonder what kind of service the tractor has had," Bashford says. "Check the coolant, too." The condition and regular changing of filters, hydraulic hoses and fluid and grease also tell a lot about a tractor.

Sun-bleached or otherwise flat paint is usually a dead giveaway that the tractor has been left in the field at the mercy of elements. A new paint job is a mixed blessing. It adds value to a tractor, but may also hide oil leaks, evidence of which was steamed off before painting.

Tires

The difference between good rear tires and tires that won't last long can be anywhere from $500 to more than $2,000 in the value of the tractor. Rear tires can be considered good for several seasons if there are not large cracks or cuts and at least 75 percent of the tread is left.

If a tractor you like is mechanically sound, but has bad tires, check the price of new tires before deciding what you are willing to pay. Be sure that the tires already on the tractor are the proper size. Sometimes the dealer or the previous owner will substitute a good set of undersized tires just to make the sale. Even new tires that are undersized will not last long on high-horsepower tractors pulling heavy loads. For recommended tire sizes, check the operator's manual, ask the dealer to show you his company's specifications, or consult a new tractor tire dealer.

Front tire condition is critical if the tractor is going to be used with a loader or front-mounted liquid carrying tanks. Without those options, tires with a few small cracks and some tread wear can last a long time.

Brakes

With the tractor going forward, apply both brakes together to check for even holding. Turn to the right, then to the left, holding the pedal for each side down hard to see if it locks the wheel in the turn. Listen for squeaking or screeching noises that might indicate excessive wear. A closer look can be had by asking the dealer to loosen a few bolts and remove the brake covers.

Clutches And Transmission

With the engine running, depress the transmission clutch and move the shift lever through all the gears. It should shift smoothly without a lot of gear clash. Adjustment can sometimes eliminate gear clash. However, in some tractors, a little gear clash is the result of design, rather than a bad clutch.

If the dealer is agreeable, field test the tractor. "The only way to really check (the transmission and clutch) is to get into the field under load and see if it pops out of gear," Huber says.

Field tests may also reveal other serious problems. "Sometimes a tractor will burn oil, but you'll have to run it pretty hard for awhile to notice that," adds Bashford.

The PTO clutch is another item that is best evaluated under load. A chattering sound upon engaging the PTO indicates a faulty clutch. If the PTO shaft continues turning after the control is moved to OFF, there is also a problem.

If it's not possible to field test the hydraulic system, ask the dealer to attach a testing device at a remote hose outlet to see if the pump is circulating the rated gallons of oil per minute and maintaining the right pressure.

Steering

To check the steering, drive the tractor in high gear on smooth pavement. If the front wheels shimmy, the steering gears, linkage or the power steering unit may be in need of major repairs. If the wobble is not too bad, sometimes adjustment can reduce the looseness. Bad universal joints along the steering shaft may also contribute to sloppy steering. If the tractor has a special hydraulic pump for the steering, look for evidence of oil and dirt buildup around the pump and the hoses. This could tell of a bad pump or leaky hoses.

Cooling System

A good radiator will not have battered fins or show evidence of repeated repair jobs. Look closely to see if there are any signs of leaks. Use a flashlight to check the fan side, too. Check the filler pipe and hose connections. A good-looking radiator improves the chances that the tractor has not been repeatedly overheated, which can cause excessive wear on pistons, cylinders and rings, and even crack the head or block.

Engine

"You can establish the engine condition fairly well with mechanical tests," says Huber.

A dynamometer test is the best method of determining the horsepower output and fuel efficiency of the engine. A dynamometer is a machine that can put varying degrees of load on the engine while the tractor is stationary. The specifications of performance are compared with instrument readings from the test machine.

You can get accurate performance and other specifications on any of the 1,600 tractors run through the "Nebraska tractor tests" since 1920 by writing Tractor Test Lab, Agricultural Engineering Labs, University of Nebraska-Lincoln, East Campus, Lincoln, Neb., 68583-0832. Include the make, year and model number of the tractors you are interested in. Test results cost 40 cents each. A pocket-sized book listing results for 150 of the more popular used and current models is available at $1.50, plus postage.

A gasoline engine that has been overhauled with oversize pistons may deliver even more than its factory-rated horsepower. The same is true of a diesel engine customized with a turbocharger. Badly worn pistons, rings, and valves will result in drastically reduced horsepower ratings on the test. This test can also be used to adjust the engine for improved performance.

Many reputable dealers run dynamometer tests on used tractors even before they reach the lot. They do this so they will have a good idea of a tractor engine's true condition. They must decide whether to offer a tractor for sale as is, or to overhaul the engine. If the dealer says the tractor has been overhauled, you may want to ask to see the work order to learn what new parts were installed.

If no major repairs have been made to the engine, ask to see the results of the dynamometer, compression and leakage tests, if they have been made. This information should also be on a work order.

If no dynamometer test was made, or if you have any reason to doubt the test results you have been shown, ask the dealer to run a test in your presence. If he's not interested enough in the sale to do it without charge, you might offer to pay for it yourself. The cost is about $25 in some areas.

A test in your presence will let you check for oil leaks. If they are bad enough to be concerned about, they should show up after about half an hour under full load. This is also a good time to recheck the cooling system. A dirty cooling system will make the tractor run hot in that length of time. Check to see if the temperature gauge is reading in the safe operating range.

Don't overlook your own comfort when choosing a tractor on which you may spend hundreds of hours. Does the seat adjust easily? If not, ask the dealer to free up the adjustments so that it will move in all directions. What about the control levers? Are they within easy reach? Do you like the "feel" of being on the seat with your hands on the steering wheel?

Price

If the tractors you like are in good condition and are among the more popular models on the market, you probably won't have much success in haggling over price. "New machinery is priced out of sight. Used machinery is even getting expensive," says Bashford in Nebraska. Farmers aren't buying many new tractors, so dealers have fewer trade-ins.

A good source of price information is the *Official Guide to Farm Tractors and Equipment,* which is published every spring and fall by the National Farm and Power Equipment Dealers Association. It lists the average value and loan value of tractors and other farm machinery. The report costs $47, so ask your dealer to show you his copy. Some rural banks have the price guide. Prices of new and used tractors and combines can also be found in the so-called "Blue Book," *Hot Line Farm Equipment Guide's Quick Reference Guide* published yearly by Hot Line Inc., Box 1052, Fort Dodge, Iowa 50501. Cost is $20. The price of a used tractor may be even more than the tractor cost originally, which makes careful shopping even more important. Like it or not, a good used tractor will cost a bundle, so you might as well get what you pay for.

Unless you need a tractor right away, expect to spend at least a week or two shopping around. Taking even more time is better. The only risk you run is that someone else will buy one of the better tractors you have your eye on. That's the trade-off for not getting in a hurry and buying something you'll be sorry with later.

When you finally decide on *the* tractor, the dealer will ask you to sign a purchase agreement. But before you sign, make sure you read and understand everything it says. If the agreement does not spell out the warranty on used equipment, which might also include discounts on parts and labor, ask the dealer to write it on the agreement. Any other verbal agreements for optional equipment should also be put in writing.

This system is not foolproof, but in 15 years of farming it has helped me buy three *good* used tractors. I'm still using them all. The system is not limited to tractors, either. I used the same procedure in searching for a used pickup truck. The search took nearly one year, but I got more truck for a lot less money.

S. E. Durcholz farms 160 acres near Jasper, Ind.

'Weatherproof' Your Farm With Irrigation

Water is the best fertilizer there is. That's why every component of your diversified small farm should be irrigated. Whether you flood whole fields or meter out water a few drops at a time, successful irrigation depends on only one thing—an adequate source of water. It must be reliable and provide water that is clean and of good quality.

On the average, a minimum of 8 gallons of water per minute (gpm) is needed for each acre to be irrigated, especially if the system is to be run 24 hours a day during periods of drought. This means that irrigating just 10 acres requires a water source capable of delivering 80 gpm. That same rule of thumb holds true for both wells and small, flowing streams used to supply irrigation water.

Ponds may also be used to supply irrigation water. A good rule of thumb for such surface water supplies in humid regions is that 1.5 acre feet of water (the amount it would take to cover 1 acre with 1.5 feet of water) should be stored in a reservoir for every acre that is to be irrigated. Irrigating 10 acres from a pond, then, would require a 1.5-acre pond that contains 10 feet of water.

In some cases, municipal or community water systems may be used, particularly with drip or trickle irrigation systems. Commercial water rates may be available to farmers using water from such a source, especially if irrigating is done during off-peak hours, when demand for water is lowest. Buying water is still fairly expensive, but it may prove to be cheaper and more reliable than digging a well or a pond.

After you've assured yourself of an adequate water supply, the next step is to choose the irrigation method that best suits your needs. That calls for the assistance of a consulting engineer, a competent irrigation system dealer or other experienced individual, since proper design is essential to having a successful irrigation system. Your Extension office can refer you to local irrigation experts, as well as give you helpful bulletins and reports that explain various irrigation techniques. The following is a quick overview of drip (or trickle) and sprinkler irrigation systems—the only kinds of irrigation worth considering for a small, diversified farm specializing in high-value crops.

Drip/Trickle Irrigation

Drip or trickle irrigation is a method of slowly applying small amounts of water directly to the plant root zone. This system greatly reduces water consumption and, if managed properly, eliminates moisture stress by continuously maintaining favorable soil moisture conditions.

It is a versatile type of irrigation that can easily be used on very small plots or extremely large plots of everything from citrus, peaches, apples, avocados, grapes, and ornamentals to high-value vegetable crops such as staked tomatoes and strawberries.

The basic operating characteristics of drip irrigation are simple. Water is pumped from a water source through a pipe network consisting of small-diameter plastic tubing that runs down each row. Water is supplied to each plant by outlets called emitters that are located along the tubing.

Like any type of irrigation, drip has certain advantages and disadvantages that should be carefully considered.

Advantages

- Requires less water.
- Uses lower pressure and less energy for pumping. Also, the pump and pipe network required can be smaller.
- Allows application of water precisely where and when it is needed.
- Minimizes water waste.
- Reduces labor needs.
- Reduces disease and insect damage, because foliage is not wetted.
- Reduces weed growth between rows.
- Allows normal field operations to continue without interruption, since only a limited area around each plant is wetted.
- Fertilizes and irrigates at the same time by applying fertilizer through irrigation lines. This often *reduces* fertilizer needs, since water is applied only near crop plants.
- Eliminates run-off on hilly terrain.

Disadvantages

- Emitters are small and may clog with sand, silt, clay, insoluble chemicals, and biological materials such as algae. Filtration is required.
- Moisture distribution in the soil is limited, sometimes requiring daily irrigation to maintain a high level of soil moisture.
- May be damaged by rodents, insects and laborers.
- Requires a high level of management.
- Does not generally provide frost protection.
- Costs more than sprinkler systems.

A drip irrigation system consists of the following components, each of which must be carefully selected to ensure satisfactory operation:

Emitters: Discharge water at very low flow rates and often at low pressure. There are two basic types of emitters: line-source emitters and point-source emitters.

Line-source (continuous) emitters deliver water continuously along a row and can be used for irrigating crops such as staked tomatoes, strawberries and other closely spaced row crops. Line-source emitters are available in several different configurations, but basically consist of small-diameter tubing with small orifices located at variable

Large, self-propelled irrigation rigs (above) may look grand, but the humble drip/trickle irrigation line (below) does a much better job with diversified, high-value crops.

Trickle irrigation lines are barely visible, but the proof of their effectiveness is easily seen in the vigor of these blueberry plants.

spacings along the tubing. Sometimes, line-source emitters are placed on the surface of the soil, but more often, they are buried 1 to 2 inches in the soil.

Point-source emitters (individual), are attached as needed to small-diameter tubing and discharge water to individual plants. This type of emitter is normally used for crops such as citrus, peaches, apples, pecans, grapes and other tree and vine crops. The number of emitters per plant is based on the daily water needs of the plant.

Distribution Systems: Typically, half-inch or three-quarter-inch polyethylene (PE) tubing is used. It is important that a high-quality tubing be used, because poor-quality black plastic tubing is subject to cracking. Buried polyvinyl chloride (PVC) mains and submains are often used to deliver water from the pump to driplines. Surface tubing may also be used for submains.

Filters: Equipment includes sand filters, screen filters, cartridge filters and sand separators. In some cases, settling basins may be used to allow certain particles and mineral materials to precipitate from the water. Whatever the filtering system, it must be carefully chosen and maintained.

Pump Power Unit: After the flow rate and operating pressure of a drip irrigation system are determined, a pump and power unit must be selected that will deliver water at the desired flow rate and pressure.

Optional Equipment: Can include timer, tensiometer to sense soil moisture and turn the system on and off automatically, chemical injector, flow meters, pressure regulators, and flush valves.

Sprinkler Irrigation, Solid Set and Portable Systems

A sprinkler irrigation system consists of a pipe network with attached sprinklers that spray water over the soil surface. Sprinkler systems are widely used in the United States for irrigating a variety of crops. Their use has rapidly expanded in recent years.

There are many different types and classifications of sprinkler systems; some have applications on small farms, while others are designed for huge field areas.

The simplest sprinkler system is referred to as a portable system and consists of a minimum amount of portable aluminum pipe with sprinkler risers located along the pipe. Typically, a portable system is used to irrigate one area, then is moved to irrigate an adjacent area. This procedure continues until the original area again requires irrigation. The system is then moved back to the original area and the process is repeated. Obviously, this type of system is highly labor-intensive. Low initial cost sometimes makes the portable system appear attractive, but difficulty with labor scheduling and water management often negates this advantage.

An adaptation of the portable system is the solid set system, which consists of sufficient pipe and sprinklers to irrigate an entire field *without* moving the pipe or sprinklers. The pipe network may consist of buried PVC pipe or portable aluminum pipe.

A solid set system consisting of buried PVC pipe and fixed sprinklers is called permanent. This system has many labor and management advantages. Its permanent risers can get in the way of field operations, though.

Another adaptation of the solid set system is a semi-permanent system, which consists of underground pipe with risers located throughout the field. The risers have quick connecting couplers that allow quick attachment of sprinklers. Sprinklers are then moved from one riser to another. This minimizes the number of sprinklers required and reduces the cost of the solid set system.

A solid set system consisting of portable aluminum pipe is called solid set portable. It is normally placed in the field for the entire growing season and then is removed for field operations or for use at another location. Movement to other locations is often desirable for crop rotation and multiple use of the system.

Proper design of a sprinkler system is essential for satisfactory operation. Sprinkler selection is an important consideration in design of any sprinkler system. Sprinklers should be selected to apply water at a rate allowing the soil to absorb water without runoff, and should be spaced so that water is applied uniformly over the irrigated area rather than wetting some areas more than other areas.

Sprinkler spacing will depend upon the sprinkler selected. Low-gallonage, medium-pressure sprinklers are often spaced on 40-, 60- or 80-foot centers, while medium-gallonage, medium-pressure sprinklers are spaced on up to 100-foot centers. High-gallonage, high-pressure gun sprinklers may be spaced more than 160 feet apart. In most small farm operations, low- or medium-gallonage sprinklers have advantages over high-flow-rate, high-pressure sprinklers.

Don't Guess, Soil Test

It's the only way you'll ever know the true fertility of your soil.

"When in doubt, follow the directions," is usually pretty sound advice. In the case of soil sampling, it's not enough. Even if you're not in doubt, follow the directions exactly. Don't assume that the way you've sampled in the past is right. Sampling methods and handling may differ for different labs, regions or nutrients.

How sampling affects the chemistry and results of soil analysis is beyond the scope of this book, but suffice it to say that the reliability of the information provided by the lab depends on correct sampling. "And no information is better than bad information," says Dr. William C. Liebhardt, director of the Sustainable Agriculture Program at the University of California-Davis.

Here are some general sampling guidelines to supplement the information provided by your lab. Additional sampling information and advice are available from your local Extension agent or land grant university soil testing lab.

Plan ahead: Don't wait until the last minute. Write to soil testing labs for information now. Consult your Extension agent for a list of labs serving your area, and information on proper sampling procedures for your region. File the information away until you have a chance to sit down and study it thoroughly.

Familiarize yourself with the services offered. Do the private labs base their recommendations on land grant research? Find out how long it will take to get test results, and what special sampling and handling procedures may be necessary. Compare costs. If the material provided doesn't answer all your questions, write or phone the lab.

Number of samples: As a general guideline, 15 to 20 cores mixed together to make a composite sample can usually adequately represent a 5- to 8-acre area. In some areas with more uniform soils, this number of cores can represent a larger area. Check with your Extension agent for local guidelines. Care must be taken to avoid areas that aren't representative of the field as a whole. If your're the only one who knows where the small, poorly-drained spot in the field is, you're also the only person who'll know not to sample there.

Handling and drying: Again, follow lab directions. Some tests require air-dried samples. Others need wet samples. Some nitrate tests require cooling or freezing. No matter how unusual the directions may seem, follow them, or the information provided may be unreliable.

Check your work: Make it a practice to "test your soil tests." If you're not sure you're sampling and testing small enough areas, divide fields based on soil type, drainage, slope or other fertility gradient and sample each part separately. If the test results are very different, you should sample and manage each part of the field separately.

Even if you've come to settle on one lab that seems to meet your needs quite well, split your samples occasionally and send them to different labs as well as your favorite, and compare results.

The right tools: Soil sampling can literally be a pain in the back—especially if you don't have the right tools. At the very least, you should have a good-quality soil probe. They're available at farm supply stores and through mail order suppliers. Like any tool, they vary in quality and price (from about $30 to $200). If you don't want to invest in one right now, many Extension offices have probes they rent or lend, or you may be able to borrow one from a neighbor.

Chapter 5
High-Value Crops And Enterprises

Being Profitable Means Being Different
And those differences guarantee this farmer a year-round income.

OPELIKA, Ala. — Back in the mid-'70s, when he first heard about the Whatley plan for making $100,000 from 25 acres, Frank Randle probably had a sudden urge to tell his boss to "take this job and . . . " Randle was working as a state bee inspector and farming part-time. Despite his distaste for bureaucracy, paperwork, and being on the road a lot, Randle kept his state job. But he also kept working on his 40-acre farm near here, adding beehives, rabbiteye blueberries, drip irrigation lines, muscadine grapes, fencing and sheep, trellising and kiwifruit until he had everything ready.

Finally, in 1982 — after nine years and six months with the state (but who's counting?) — Randle told the boss he wasn't working there anymore. He was going to farm full-time.

At first, Randle says, not picking up a paycheck every other week was traumatic. But then all the hard work he had put into his farm began to pay off. And just two years later, Randle found himself featured in *The Wall Street Journal,* which reported that he was earning more than twice the $20,000 a year he made working for the state.

"Mr. Randle says he has a neighbor just down the road with a big soybean farm who is $750,000 in debt and may not be able to make the next interest payment. 'I don't owe anybody anything,' " the article quoted Randle as saying. Randle was cited in the article as a shining example of a farmer successfully following part of my small-farm plan.

Diversity Means Security

"We're not getting rich. We have a fairly comfortable lifestyle. But we are paid out," the 35-year-old Randle says today.

"We're what you might call peripherally involved in the Whatley plan," he adds. Randle has the extensive fruit and berry plantings that I prescribe, plus a thriving apiary and sheep operation. There is just no time, though, for vegetables or even strawberries, which he found required too much herbicide and fungicide spraying. "I am spread so thin with what I am doing now, that I am hesitant to take on any more. I am sure those other things like vegetables will work, but my limitation is time. We have been successful with what we have tried so far." And Randle means to keep it that way.

HIGH-VALUE CROPS AND ENTERPRISES

Frank Randle and son Zachary, 3, survey sheep pasture after a morning of fixing fence. He often takes care of Zachary and his other son, 5-year-old Franklin, while his wife, Pat, works as a registered nurse.

Randle's main enterprise is, not surprisingly, honeybees. He began raising bees at the age of 8 when a beekeeping relative gave him a beehive, and now has 1,100 hives. Each hive produces more than 100 pounds of honey per year, which he says is nearly 2½ times the state average for 1986. Some of his hives even yielded about 300 pounds that year. "That's a full-time job right there."

True to the Whatley formula for success, though, Randle adds, "We're diversified in everything we do." With the bees, for example, he does not just sell honey, beeswax and bee pollen. Randle also rents his hives to fruit and vegetable growers who want to improve pollination. His hives are scattered around some 40 locations in Lee County. He sells queen bees and a variety of apiary supplies to beekeepers throughout the United States. In fact, he even entered the export business recently, selling packages of bees to the United Arab Emirates for pollination in citrus groves. "We're always looking for new markets. I never thought our customers would be as far away as Abu Dhabi, though," he grins.

"Last year wasn't too great with the drought, but I can't complain," Randle says. "I had the best honey crop I've ever had."

Bees Boost Yields

Proper pollination is essential to achieving the yields necessary to earn at least $3,000 per acre. That is why I tell farmers they should have 60 beehives. Actually, at our test farm at Tuskegee we did much better than that on income. Our production of berries and sweet potatoes was enough to bring in nearly $8,000 per acre for some things, more for others.

We put out 605 blueberry plants per acre. Each plant produced about 25 pounds. If we sold them for just 60 cents a pound — and you can get a lot more than that — we would have made over $9,000 per acre. If we can do it, then so can the farmer on his own land.

With blueberries, there is a trick to guaranteeing good pollination. You want to make sure the bees pollinate every single blossom on every single plant. So you want to fool the bees. Move your bees in a day or two before your blueberries come into full blossom, and shut them up in the hive for the night. Pick yourself a handful of blueberry blossoms from each variety, and boil them in a pan of water.

Let it cool and then pour about half a teacup of that water at the entrance of each hive. The bees will smell that aroma all night long. And when you let them out in the morning, they'll head right for those blueberries.

Berries, for the most part, have not been a major crop in Alabama. But we can make it a major crop. Down here, we buy most of our blueberries from New Jersey. But we can grow them just as well in Alabama and have them on the market three weeks ahead of New Jersey.

There is already a thriving regional market for seasonal delicacies like fresh blueberries. Lately, more and more people are driving the 60 to 75 miles from Montgomery and even Birmingham to pick their own berries at Randle Farms. "I let the Boy Scouts and school groups come out and pick for free, because then mama comes out to the farm. There's a madness to my method," Randle chuckles.

He sells pick-your-own blueberries for 75 cents to $1 a pint, which is about half of what they bring in area groceries. About 1.5 pints make up 1 pound of berries. In 1986, which was a good year for blueberies, Randle says his 4.5 acres of blueberries yielded about 12,000 pounds of fruit per acre.

For customers who can't make the trek out to the farm when the blueberries are ripe, Randle fills two freezers with about 1 ton of berries. Those customers include the growing number of yogurt shops in the area. One such shop even worked with Randle to develop a syrupy blueberry topping for its yogurt. The shop now buys 5 to 7 gallons of topping each week at $12 per gallon.

Randle has 605 blueberry plants per acre. They are spaced 6 feet apart in the rows, which are 12 feet apart to aid mowing and harvesting. Drip irrigation lines water all of his blueberries and other fruit plantings. When he started farming, Randle fertilized with rabbit manure from the rabbits he was raising for sale as lab animals. As that market dried up, he began raising sheep and now fertilizes with their droppings. The only purchased fertilizer he uses on the blueberries is about 200 pounds of ammonium sulfate—spread out over all 4.5 acres of blueberries—each spring. "Next year's crop is on this year's growth," Randle explains. "That gives me 6 to 8 inches of growth real quickly, and assures me of a crop next year."

With all the pickers and other activity in his berry plantings, Randle says he doesn't consider birds to be much of a problem. "We don't worry about it. I don't mind feeding a few birds."

Problems Ahead?

"We've made good money with diversification, very good money," Randle says. "But now Extension has discovered diversification and is promoting it. Price is starting to go down. For example, Extension is promoting kiwifruit as one of the solutions. They want 100 acres of kiwi in the area. You can have too many people doing the same thing, whether it's growing cotton or corn or blueberries.

"You almost have to keep your light under a bushel, so I don't talk about what I'm doing too much. I had to cut my price this year to stay competitive. We're approaching a break-even price. There is nothing wrong with competition. I'll stand with the best of them. But I see problems coming down the pike."

On the plus side, Randle says running a highly diversified small farm with labor-intensive crops is not for everyone. "There are very few people who are willing to make that long-term commitment," he adds.

Breaking New Ground

Concern about overproduction of so-called specialty crops like kiwifruit and berries is one of the reasons that Randle Farms continues to diversify. Randle and a long-time friend, Bob Rogow, are doing what many said was impractical in the Deep South. In the early '80s, they began raising sheep commercially. They now have 200 ewes. Rogow, who also handles Randle's accounting, teaches accounting and tax management at Auburn University. Randle earned a bachelor's degree in entomology at Auburn and did graduate studies there in agronomy.

"We started out knowing absolutely nothing about it. There is no tradition of sheep in the South. It was something we felt we had to do. The country's population is shifting southward and so are Northern and Western tastes for sheep," Randle says. "We're seeing an increasing demand for sheep in the South. If we don't try meeting this demand, we'll be missing a real opportunity. The sheep fit into things really well. They graze the woodland tracts and even eat the kudzu." The lambing rate for his Dorset ewes is 186 percent. But, when coupled with those of his crossbred Suffolk and Rambouillet ewes, his overall lambing rate drops to 137.5 percent. "Can you guess what kind of ewes I'm going to be getting more of?" he says.

Computer Helps

Randle has computerized all of his production records. He uses a Corona personal computer that cost about $2,000, including a printer and other hardware. Despite problems finding the software he needs, Randle says the computer is a big help. "It sure is a good way to track customers and do mailouts, keep lists and keep production records, especially with the sheep." Although Randle has many loyal customers, he has yet to form a Clientele Membership Club and charge an annual membership fee, payable the first of the year. "It's not a bad idea, though. It gives you your operating money at a good time."

But when it comes to marketing, Randle takes to heart my advice about eliminating middlemen. He'll have nothing to do with livestock auctions. All his lambs are sold to individuals, priced on live weight. He delivers the lambs to a local slaughterhouse where they are killed, cut to the customers' specifications and wrapped. Most of the lambs go right into customers' freezers. Slaughter weight ranges from 100 to 125 pounds. Price is about 80 cents a pound. Randle gets more than double that price—$1.75 a pound—for the 30- to 40-pound "Easter lambs" he sells to Greeks and members of other ethnic groups each spring.

Besides slaughter lambs, Randle sells replacement ewes, show animals, wool and high-tensile, electric fencing and supplies. Sheep are the perfect livestock enterprise for smaller-scale farmers, Randle says. He and Rogow are doing their part to promote and improve sheep production and marketing in the their state. They recently founded the Alabama Sheep and Wool Growers Association.

"There is a real high demand for lambs that aren't pumped full of antibiotics and growth-promoting hormones," he explains. In marketing all of his farm products, Randle stresses the fact that he uses few, if any, chemicals. "We have customers come here from within a 100-mile radius, Birmingham, Columbus and Atlanta, Ga., and into Florida. Competition is getting tougher, but we have the reputation—and the clientele. There are a lot of people raising sheep between here and Florida, but the customers come here. People just seem like they're really concerned about where their food comes from."

Diversity Breeds Diversity

Randle worries less about increased competition among sheep producers, because of the wide diversity that's possible in the sheep business, itself. One option he's seriously considering is adding colored sheep to his flock to tap the growing demand for specialty wools. Also, all sheep producers need fencing. Randle will happily supply them with the high-tensile, electric fencing that he sells and also uses on his farm.

HIGH-VALUE CROPS AND ENTERPRISES

"There is not a piece of barbed wire left on the place," Randle says. "It's all high-tensile, electric. It gives us excellent animal control and it doesn't cost all that much. I can't say enough good things about it, and that's not just because I sell it."

Randle uses 5- to 7-wire perimeter fencing and divides his pastures with 3-wire fencing. "Knock wood, we have not lost a lamb or a ewe to any predators," he says. The fencing keeps dogs and other predators out. But, more importantly, since much of Randle's pasture lies beside a busy highway, it keeps the sheep in.

And Randle really packs his sheep into the pastures. "I'm doing what's called mob grazing, putting 200 sheep on 1½ to 2 acres of grass," he explains. "When the first few sheep start to lie down, I kick them off, back to the drylot. That deposits a tremendous amount of manure in one place. I am starting to eliminate fertilizer, just using a little lime. With a little rain, I've got grass that grows like you won't believe. It's so dark green it's almost black in places. Weeds just ceased to exist. There is no weed problem . . . only extremely uniform stands of pasture grass." Randle currently rotates grazing through 21 paddocks, 16 of which are on his own land. "I may double that (latter) number. Thirty-two is supposed to be the optimum number, because then the pastures rest 94 percent of the time."

On his hay fields, Randle is still using some fertilizer, but says that the fertilizer recommendations from his soil test lab have gradually been declining over the past six years. He wants to reduce the need for purchased fertilizer even more. Toward that end, Randle just planted 20 acres to TIBBEE crimson clover, which he will harvest for seed to plant on all of the hay ground he owns and rents. A good seed yield is about 200 pounds per acre. What seed Randle doesn't use, he can sell. A local seed dealer has already offered to buy as much seed as Randle wants to sell for 65 cents a pound. The going retail rate in the area is 85 cents. Not a bad proposition, Randle says, considering that establishing the clover only cost him $10 an acre for seed, plus tractor time.

'Lead, Don't Follow'

All that is quite a change from the condition Randle's land was in when he bought the farm in 1975. The land was all in corn then, with rows running straight up and down the rolling terrain. Erosion was a real problem. His land is now tucked securely under a protective blanket of forage grasses and legumes. Long rows of fruit and berry bushes and vines follow the contours of the slopes.

"The fruit, sheep and bees complement each other, they don't compete," Randle says. "All of the things I'm doing fit in perfectly with the bees. That's the key to this sort of farming. By being diversified, I'm assured a year-round income. Farmers need something to sell every day.

"The whole emphasis in farming has been to get bigger. Our farm is an example of how wrong that idea can be. The small, family-owned and -operated farm like ours is the backbone of this country. It always has been and I think it always will be. There are still a lot of opportunities in agriculture. But you have to be a little off the wall, sometimes, and kind of a visionary, like Dr. Whatley. We've been successful because we're different. You have to lead, not follow.

"But you have to stay right on top of it all," Randle cautions. "We're limited only by time, not ambition."

These Little Lambs Mean Big Profits

Feeder lambs can net about $4,000 per acre per year through super livestock and forage management.

Small farmers cannot afford the luxury of maintaining a flock of ewes and rams year-round. When you get right down to cases, it just costs too much to feed all those animals, house and medicate them, and guard against predators all year long. There's no need for all that, because there is a better way. That is to *buy* feeder lambs from farmers around the country who specialize in raising feeder lambs. Let those farmers do the lambing, castrating, medicating and guard-dogging for you. Of course, you'll pay for all that when you buy their lambs, it's true. But, in the long run, you'll save even more in time, money and management headaches by being freed of the burden of maintaining a flock year-round.

To make a feeder lamb operation truly successful, *the first thing you need is a guaranteed market*. That means you need a Clientele Membership Club (CMC) associated with your farm. If you don't have a CMC, then you can just forget about raising feeder lambs, and I mean right now. Skip the rest of this section and go on to something else.

The other thing you absolutely, positively must have to be successful with feeder lambs is a USDA-approved slaughterhouse within a reasonable distance of your farm. Ideally, the slaughterhouse should be no more than 10 to 15 miles away. A lot of smaller slaughterhouses have been closed in recent years, so I guess you could go up to 40 miles away, but that's really pushing the limit. The distance to the slaughterhouse is critical, because you'll not only be hauling the lambs there for slaughter, but you'll also be bringing the processed meat back to your farm under refrigeration, many times each year.

Everything has to be just right with this kind of operation. Draw some hard and fast management guidelines — and stick to them. There is no point — or profit — in bending the rules of common sense and getting in way over your head with outrageous transportation and other costs. That's why it is important to keep in mind that the plan outlined here is *only a model* for a 300-head feeder lamb operation in the Deep South. It can and should be scaled up — or down — and otherwise adapted to perfectly fit your particular location and circumstances.

A feeder lamb operation is a high-value component that should provide a NET return of more than $4,000 per acre per year in five years after you amortize fencing, solid set irrigation system, freezer and refrigeration facilities and receiving barn. Lambs may be fed out following this program in all 50 states of the United States, and in developed countries around the world. Only the number of feeding cycles will vary, from three in the Deep South where you have at least a 180-day growing season, to one in the far northern United States and southern Canada. What this component requires is high-level management of both lambs and alfalfa in order to get optimum average daily weight gains from the lambs and maximum alfalfa production. After you feed out your last lambs each year, the alfalfa should make sufficient regrowth to provide adequate cover for you to operate a game bird hunting preserve during the fall and winter months.

Now, let's assume that you have a slaughterhouse nearby and also that up to 300 of your CMC members want lamb. All have signed contracts with you stating that, starting next year, they will each purchase from you a 120-pound (live weight) lamb for $1.50 per pound. For that price, the lamb will be slaughtered, aged, cut to their specifications and wrapped for their freezer. Each lamb will dress out at about 60 pounds of meat. You should emphasize to your customers that your lambs have not been injected with estrogen or other growth stimulants.

If you mean business — and only if you mean business — adapt this program to your local conditions:

FIRST YEAR

1. Tell the manager of the slaughterhouse near you that, starting next year, you will have one, two or three groups of 120-pound lambs to be slaughtered in groups of 100 at roughly 64-day intervals. They will need to be slaughtered, aged, custom cut, wrapped for the freezer and individually boxed. The slaughterhouse's fee for this will be about $15 per head. Get a quote on their price — in writing. Also, specify that, in addition to cut and wrapped meat, your clients are to receive the liver, brains, head and blood, if desired. The slaughterhouse keeps the hides and trimmings.

Running a traditional flock of ewes and rams year-round isn't going to make as much money for you as stocking feeder lambs.

2. Identify sources of feeder lambs in California, Colorado, Missouri, Texas, Virginia and Wyoming. Use established suppliers! As work progresses on your feeder lamb pasture, consummate your agreement with lamb suppliers early — and in writing. Your contract should stipulate the ages, weights and sex of the lambs to be delivered (your feeder lambs should all be castrated males, which are known as "wethers"), plus vaccination records.

3. Select a level, well-drained 10-acre site on your farm that has never flooded. Weed control on this site must be excellent, since you want a pure alfalfa stand without herbicides.

4. In February, contact your county Extension agent to determine the recommended varieties of alfalfa for grazing rather than cutting, seeding rate and approximate planting date. Locate a reliable seed source.

5. In March or April, take four random, composite soil samples and submit them to your state's land grant university soil testing lab or a private lab that is familiar with your soils. Tell them that you will be planting this 10 acres to irrigated alfalfa for intensive, rotational grazing. Ask for recommendations for lime to adjust the pH to 6.8 to 7, and for phosphate and potash fertilizer.

6. Actual field work begins in April or May. Subsoil, if necessary, plow and disk your 10-acre site. Lime and fertilize according to soil test recommendations. Plant a summer legume such as soybeans or cowpeas for green manure. Be sure to inoculate the seed with *Rhizobium* bacteria.

HIGH-VALUE CROPS AND ENTERPRISES

7. Contact an agricultural engineer at your state land grant college for help in designing a solid set irrigation system. The purpose of this system is to "weatherproof" your paddocks, since regrowth varies with weather and you must be sure to maintain optimum alfalfa production.

The engineer will need to know that you will be using high-tensile, electric fencing (usually 5-strand) for perimeter fence, and that the 10-acre field will be divided into eight paddocks consisting of 1.25 acres each. Fencing for paddocks can consist of 3-wire electric fencing or even temporary fencing to allow you greater flexibility in adjusting your grazing rotation.

You will want a centrally located source of water that is available to all paddocks. Be sure to mention that the alfalfa will be grazed up to seven growing seasons, depending on your location. Afterward, the field will be planted to sweet corn and other high-value, annual crops. You will be feeding lambs 1.5 pounds of a complete high-energy, high-fiber ration of corn, barley and oats per head daily, plus salt and Bloat Guard (poloxalene). Go heavy on the barley and oats. Mississippi State University reported recently that beef and poultry fed on barley and oats had lower cholesterol levels. You will need four portable feeders.

8. Contact suppliers of high-tensile, electric fencing and pick your brand after comparing prices, specifications, local availability, service and other features. You can install this fence yourself. It's not all that difficult. Most suppliers provide excellent directions. And besides, you can save a lot of money—and make sure the job is done right.

Traditional electric fencing has developed a very bad image, because it requires a lot of maintenance, like constantly cutting weeds and grasses from under fencelines to keep them from grounding out. The new, low-impedance fence energizers are so high-powered and well-designed that they will not ground out when fencelines come in contact with grasses and weeds. You will still want to control weeds under the fence, though, to keep your farm looking pretty for your CMC customers. And remember, a lot of those customers are coming to you because they want "chemical-free" food. You want to keep them happy, so figure on using a gasoline-powered weed trimmer or a good scythe, rather than herbicides.

This new fencing is often called New Zealand fence, because it was developed Down Under by sheepmen who needed low-maintenance fence that would still keep sheep in and predators out. Such fence is widely used against predators like coyotes, dogs and foxes, although its effectiveness varies. One way to improve predator control is to determine your predator risk, beforehand. For example, suburban dogs present a different challenge from that posed by coyotes. Guard dogs or guard donkeys that are sometimes used in year-round sheep operations do not really fit this system, since the feeder lambs will only be on your farm for a couple of months.

9. Aug. 1—Disk down and incorporate green manure crop.

10. Aug. 15—Plant alfalfa at the recommended seeding rate for your area. Be sure to inoculate the seed. Avoid seeding too deeply. Half an inch is deep enough for alfalfa. Cultipack after seeding to assure good seed/soil contact and apply 0.5 to 1 inch of irrigation water. Then irrigate as needed to prevent moisture stress.

11. Purchase a walk-in freezer large enough to hold all the lamb you may have to store at any one time, plus pheasant, quail and other processed meat animals you may need to keep frozen until they can be picked up by customers. You'll also need to get a refrigeration unit for your truck for transporting processed meat from the slaughterhouse.

SECOND YEAR

12. Delivery—Contract for your lambs to arrive at your farm about four days before the alfalfa reaches one-third bloom stage in paddock No. 1. (Paddocks 2 to 8 should reach one-third bloom stage at roughly four-day intervals. This is accomplished quite simply by cutting the alfalfa in the first paddock 28 days before the expected arrival of your lambs. Next, cut paddocks 2 to 8 at four-day intervals.)

Lambs should be weighed, numbered and held in the receiving barn for four days. That will give you time to worm them, give selenium and vitamin E injections, administer coccidiosis medications and feed Bloat Guard. It will give the lambs time to recover from their cross-country trip before they head for the pasture.

Be prepared to feed and water the lambs while they are in the barn. Allow 10 square feet of floor space per lamb. For 100 lambs, you will need 1,000 square feet, which means a barn that measures 40 by 25 feet. The barn doesn't have to be anything fancy. A roof with mostly open sides is fine for the lambs. For ease of handling 100 lambs at a time, the barn should be located as close as possible to paddock No. 1. When not being used for lambs, this barn becomes a perfect place to store hay, machinery and other items. It can also be used for special promotions like a "haunted barn" to delight your customers around Halloween.

13. Put lambs on pasture. Start with paddock No. 1 and rotate them to the next paddock every four days.

Having so many animals in such a small space for so long is not as damaging as it might seem. In fact, the "herd or flock effect" of this "mob grazing" is about the best thing that could happen to your pasture. For example, hoof action breaks down dry matter and eliminates soil crusting, which enhances water penetration. It also recycles manure and urine, pumping large amounts of nitrogen and other nutrients back into the soil.

Rotational grazing is part of a good worm-control program and actually lessens the need for chemical wormers. A 3- to 4-week pasture regrowth period, 28 days in this instance, is sufficient time to significantly reduce, although not completely eliminate, the number of roundworm larvae in the pasture.

14. After 64 days on alfalfa pasture and grain, with plenty of water, salt, Bloat Guard and TLC, the lambs should have reached the ideal market weight of 120 pounds each. It is time to haul them to slaughter. (Lambs not yet at market weight may be held for slaughter later. Sending fewer lambs to slaughter at one time considerably eases the pressure on you and the slaughterhouse.) A week to 10 days after slaughter, you should return to the slaughterhouse and pick up your clients' lamb, which is all custom-cut, freezer-wrapped, and individually packaged in cardboard boxes. Haul the lamb back to your farm in your refrigerated truck and place it in your walk-in freezer until it is picked up by customers.

You should already have given your CMC members the date and time they could pick up their lamb, as well as a reminder that they should each bring you a check for $180. It might *seem* easier to have customers pick up their lamb at the slaughterhouse. But don't forget the psychology behind the CMC— *you want the customers to come to your farm.* They may come to pick up lamb, but they'll leave with honey, blueberries, grape juice, sweet potatoes and maybe a pheasant or two, in addition to their lamb. Never miss an opportunity to get your customers out to your farm!

15. Be sure to give your customers a few choice lamb recipes, one of which they can use for dinner that very night. Don't forget to pass along some serving suggestions to help make sure that your customers get the most enjoyment out of their lamb. Tell folks to be sure to use heated serving platters and to never serve cold drinks with lamb. You want your customers to enjoy their lamb, and to come back to you for more, again and again. They may not be as inclined to do that if they take a bite of lamb, then a sip of iced tea and the lamb fat suddenly congeals in their mouths. So tell them to stick with hot drinks. And—before they leave your farm with their box of lamb—get their tentative commitment for buying one of next year's lambs.

Feeder Lamb Inputs & Costs For Start-Up Year

300 60-pound Feeder Lambs @ 75 cents/lb.	$13,500
Transportation to your Farm @ $5/head	1,500
Medication @ $5.00/head	1,500
Complete Ration, 1.5 lbs/head/day @ 10 cents/lb., 102 lbs.	3,060
Transportation to Slaughterhouse @ $2/head	600
Slaughtering, processing @ $15/head	4,500
Transportation from Slaughterhouse to Farm @ $2/head	600
30 tons, Dolomite Lime @ $25/ton (every 3 to 4 years, as indicated by soil test)	750
15 tons Phosphate and Potash Fertilizer/Boron @ $180/ton	2,700
Amortization: Fencing, Receiving Barn, Freezers, Alfalfa, Irrigation $50,000/amortized over a 5-year period	10,000
Pesticide, Alfalfa Weevil	150
Weed Control Under Power Fence	150
Soil Testing, 4 samples @ $5/each	20
TOTAL	$39,030

Your Own Hunting Preserve

Fact A: We have deer running out of our ears in many parts of this country—white tailed deer, mule deer, you name it. They are becoming a real nuisance, inflicting heavy damage on many agricultural crops. Farmers are allowed to kill deer that are destroying crops. But once the offending deer have been shot, there is nothing farmers can legally do with the meat. Most states do not allow the slaughter and sale of venison.

Fact B: There is a large and growing consumer demand for venison in the United States. In 1986, this country imported more than 400 tons of venison worth some $3 million from New Zealand and Australia.

Question: Are we Americans just plain stupid?

Until the production and sale of venison is legalized throughout the United States, the only legal avenue open to farmers everywhere is the sale of hunting rights. That is why I am recommending that farmers with enough land establish 200 acres of woods and fields as hunting preserves where their clients can hunt fallow deer, squirrels and doves. That allows farmers to provide dressing, cutting, aging and wrapping services without running afoul of state and other food inspectors. To sell hunting rights, farmers just need permission from their state's wildlife or natural resources department. And, of course, hunters must have valid hunting licenses and observe the normal hunting seasons.

HIGH-VALUE CROPS AND ENTERPRISES

This component is similar to the feeder lamb component described earlier in that you must have a Clientele Membership Club if you are to succeed. This is a high-value component which requires a relatively high capital outlay, initially. Look at the money involved as an investment that will yield a high return, rather than as an expense. This procedure does not solve the problem of venison imports, nor does it ease our balance of payment problems. I hope that with time, we will come to our senses, and make venison available to the general public, and that small farmers across the country can make fresh, custom cut and freezer wrapped venison available to their CMC members just as they now do with lamb.

I recommend stocking fallow deer in a hunting preserve because it is one-half to two-thirds the size of red deer and its meat is of a superior quality. The stocking rate of eight per acre is twice that of red deer and carcass weight is normally 57 percent to 58 percent of live weight. Fallow deer have a higher reproductive rate than red deer. They are also relatively resistant to tuberculosis and are not known for wallowing in drinking ponds. Also, animal lovers will not accuse you of fencing in "wild" animals because fallow deer have been raised in captivity for many years, especially in European parks where they have long been used to keep parks well-groomed. They are excellent at clearing brush without harming larger trees.

To establish a fallow deer, squirrel and dove hunting preserve, you will need to enclose 200 acres with double fencing. Perimeter fence should be 8-foot chain link fence topped with an overhang bearing three strands of barbed wire. Interior fence, located 50 feet inside of the perimeter fence on all sides, should be high-tensile electric fencing. Both fences are absolutely essential to keep out predators — both the four-legged and two-legged kind — and keep your deer herd on your land.

Woods and fields must provide year-round browsing and grazing. You will also need a stocked watering pond and a feeding stand to provide hay, grain, salt, minerals and water. Be sure to add apple cider vinegar to the water at the ratio of 150 to 1 to help tenderize the meat.

Woods should consist of oaks for acorn production, English walnuts, pecans, hickories and pines. Depending on your location, plots of brambles, honeysuckle, bamboo and sumac should be strategically planted and spaced to provide abundant ground cover and year-round browsing.

Grazing areas should provide year-round grazing on crops like alfalfa and clover, sunflowers, millet, corn, soybeans, cotton and peanuts. Winter-grazing crops will be skewed toward barley and oats, since a diet heavy in those grains has been reported to reduce cholesterol levels in meat animals. Areas will be seeded to centipede or orchard grass and overseeded in early to mid-September with annual ryegrass to provide winter grazing.

The farmer should consumate a contract with each sportsman, providing a money-back guarantee that the hunter will get a deer kill. The fee for that should be $500 to $1,000 per client per season. That price includes two nights lodging and meals. There will also be a minimal fee for dressing, aging, cutting, freezer wrapping and shipping the venison via UPS.

Deer farming in the United States and Canada is a new way of farming with excellent chances of success. It is one of the few agricultural ventures in which the demand truly exceeds the supply. As a bonus, farmers with a hunting preserve can charge an added fee for doves and squirrels. This fee would be based on a membership fee, bag limit and a service charge for any lodging, meals and processing of game. This component lends itself to total economic

Fallow deer (left photo) are easily distinguished from white-tailed deer by their short legs. On the average, fallow deer stand about three feet tall. Yet it is not uncommon for bucks to weigh about 200 pounds.

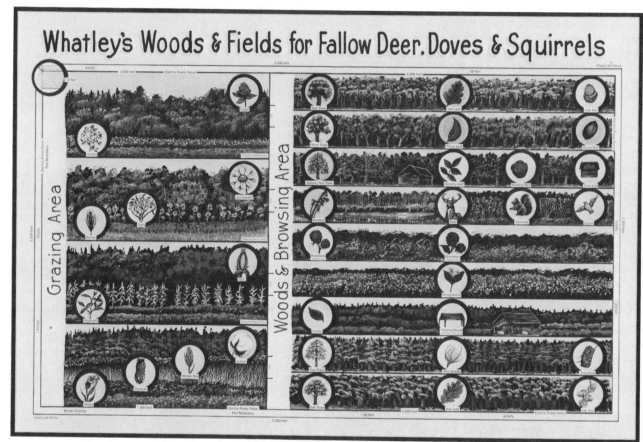

The entire area devoted to a hunting preserve containing fallow deer, doves and squirrels is ringed with both chain link and high-tensile electric fencing. To provide year-round grazing and browsing, plant such crops as alfalfa, clover, sunflowers, oats, barley, corn, soybeans, cotton, winter rye, peanuts, millet, nut trees (oak, pecan, walnut and hickory) bamboo, bramble fruit, sumac, honeysuckle and pine trees. Don't forget a stocked watering pond and feeding station.

exploitation and the value-added principle. That means taking everything from squirrel pelts and tails to deer skins, and processing them into salable items. Why, even the velvet from deer antlers is a highly sought after commodity by Orientals who use it as an aphrodisiac. The farmer should not, under any circumstances, allow others to do for them what they can and should do for themselves.

For More Information

The New Zealand Technical Correspondence Institute has developed a worthwhile correspondence course for people interested in deer farming. I recommend that you make an effort to enroll in this course, which can be started at any time of the year. Write:

Deer Farming Tutor
Agriculture Dept.
NZTCI
Private Bag
Lower Hutt, New Zealand

Even if you're not interested in setting up a hunting preserve right away, there are still many things you can do to enhance wildlife habitat on your farm, whether for hunting or the enjoyment of your customers. Contact your state wildlife department or local offices of the U.S. Fish and Wildlife Service or the Soil Conservation Service. There are also a growing number of private "wildlife managers" who specialize in working with farmers and other landowners, from governmental bodies to industrial concerns.

A good book on deer farming is "The Farming of Deer," edited by David Yerex, Agricultural Promotion Associates Ltd., 1982, Securities House, 126 The Terrace, P.O. Box 10-128, Wellington 1, New Zealand.

Seven Times Better Than Beef

JOSEF VON KERCKERINCK

Deer farming in the United States and Canada is one of the few agricultural ventures where demand exceeds supply. At Lucky Star Ranch, we started deer farming in 1978 with 50 fallow deer. During 1979, we purchased 300 more and now have about 1,100 deer, making us the largest venison producer in North America.

We raise European fallow deer because they have, by far, the best venison. They are affordable and easy to keep. Fallow deer have been kept in captivity in parks and other similar settings for more than 2,500 years, so no one can accuse us of confining wild animals.

More Demand Than Supply

Although we incurred substantial expenses and faced many problems since acquiring our first deer, the biggest obstacle was convincing would-be customers in New York City and upstate New York that we were a legitimate operation and that our venison was the best available. Once we cleared that hurdle, we faced a problem that is somewhat unusual in agriculture today. Unlike other farmers, our big problem wasn't selling our product. Our problem was producing enough venison to meet the needs of all our clients through the winter. That's one of the reasons we would like to get as many farmers as possible interested in both raising deer and joining our marketing effort. It will take years to establish a group of deer farmers that is large enough to produce even the amount of fresh and frozen venison that is currently being imported by the U.S. — 400 tons per year!

That is not a very big amount, but it represents about $3 million. Many farmers could keep their farms going if they had only a share of that income.

We made a lot of costly mistakes and are willing to help others avoid them. For example, we started out with an 8-foot chain link fence that cost a fortune. We now work with a special deer fence from New Zealand that costs about 65 cents per foot.

We have gained more knowledge about nutrition and shelter. We know that you can keep four does and their fawns per acre. For every 25 to 30 does, it is necessary to have a good, 5-year-old buck and a 2-year-old buck for breeding. We keep 600 does in one big pasture and run one buck for every 12 does, because the bucks like to fight and spend more time fighting than breeding. Rutting season begins in mid-October. Gestation period is 234 days. The first fawns are usually born around June 1.

The deer do not like to be disturbed during fawning. We only go into the pastures during fawning if there is a problem with predators like dogs and coyotes. We do everything possible to keep predators out of our paddocks. A few years ago, before we were fully aware of coyotes in our area, coyotes came and went all summer, taking 130 fawns from our pasture.

After most fawns are born, we feed grain to the does to increase milk production. That helps them raise healthy, heavy fawns that will grow into yearlings with good carcass weight. When we started raising deer, our carcass weights were in the 30-pound range. Today, through intensive culling and development of new nutritional programs, we have doubled our average carcass weight.

Fallow deer will graze on any good pasture consisting of grass and clover or a little alfalfa. They also destroy all underbrush, but don't harm larger trees. In Europe, they are used to reclaim land and keep parks clean.

Each fall, we round up all the does and fawns. Each animal is tagged and tattooed and receives injections to guard against leptospirosis, worms and other parasites. Those two injections are the only chemicals our deer receive. No antibiotics or growth stimulants are used.

Upon weaning, male fawns go into a separate pasture where we keep them until they are harvested the next fall. Slaughter normally begins about Oct. 1, when most restaurants begin to serve venison. The restaurants usually place their orders one week in advance, which gives us enough time to kill, cure and package the deer in our own slaughterhouse. We deliver the meat in a cooler unit mounted on our own pickup truck. The cooler can easily be removed after use. The idea is direct marketing, with no middlemen to take the profit that the farmer should make, himself.

Before you invest a lot of money in deer farming, the most important thing to do is to find a market. Every deer farmer has to do that on his own, unless he's lucky enough to have somebody do it for him. In some states, government agencies are a great help. Just talk to your local Extension office. We also advise prospective deer farmers to check state and local regulations and contact local wildlife and natural resources officials about any laws that might apply to deer farming.

Deer farming can be much more profitable than commercial beef cattle. A farmer who already has the land, some buildings and machinery, only needs to invest in breeding stock and fencing. A breeding doe costs $350 to $500. Bucks start at about $500, each.

Fifty does will cost about $17,500 and should produce at least 45 fawns, which can be sold at the age of 15 to 18 months for a total of $13,500. You can keep seven fallow deer on the same amount of land that one cow would require. Those seven deer will have at least six fawns, which will sell for a total of $1,500 to $1,800. A beef calf, even though it could be sold sooner, would have to be very good just to bring even 20 percent of that.

Food And Shelter

During late fall, winter and early spring, fallow deer need shelter. They cannot be kept outside all the time, because they do not like cold, windy rains or, worse yet, icy rains. If the weather is dry, they don't seem to mind temperatures as low as 40 below zero. Also, deer cannot be kept in total confinement. They need grass and room to move about. After all, they are still quite wild, being only semidomesticated. It took 8,000 years to domesticate the cow, and fallow deer have only been kept by man for about 2,500 years.

During winter, we feed our deer hay, haylage or corn silage and some grain for energy. Use whatever is available, but keep in mind that a deer needs 2.2 pounds of dry matter per day, plus fresh water at all times.

After nearly a decade in this business, we see a lot of potential for deer farming in North America. That is why we founded the North American Deer Farmers Association Inc. to serve as a forum for farmers to exchange information on deer production methods, marketing venison and other professional interests. If you have questions about deer farming, please drop me a note. Or, if you're going to be in the area, come and see our operation. Just call ahead to let us know you're coming. Here's the address:

Josef Von Kerckerinck
Lucky Star Ranch
(North American Deer Farmers Assoc.)
Box 273
Chaumont, N.Y.
13622

Phone: (315) 649-5519

Big Buck$ From Bizarre Breeds

Premium prices and reliable markets are just two reasons these farmers switched to exotic livestock.

WAYLAND, Iowa — Not long ago, Leroy and Carolyn Kaufman were still raising beef cattle and finishing 1,000 hogs a year on their 360-acre farm. Today, instead of a barn and a farrowing house, their livestock would likely feel more at home in an ark. "The experts say you've got to re-evaluate your farm, and that's exactly what we did," explains Carolyn matter-of-factly. "The hogs and cattle weren't paying for themselves, so we went strictly exotic."

The Kaufmans are one of several farm families that are using their livestock management skills to tap a steady and growing market for animals more common in zoos and game preserves than on farms. "We started in a slow way, and just added one animal at a time," says Leroy, explaining that the idea began with three white-tailed deer he bought in 1970. Since then, the operation has grown to include 80 buffalo, 40 deer, 40 elk, 14 llamas and assorted ducks, geese, emus, ostriches, peacocks, raccoons, rheas, Sicilian donkeys, swans, turkeys and zebras.

But the biggest change of all has been in their bottom line. "I'm glad we got into raising exotics when we did, the way the farm economy's been going," says Leroy. "Our markets are informal, but steady. And no one's manipulating the prices. I just sold a 6-month-old-llama doe for what I could have gotten for 40 (beef) calves."

Some of the couple's animals, like the elk, deer and buffalo, are exotic only in the sense that they haven't been domesticated. Others, like the llamas, are true, non-native creatures that are simply scarce and in demand. But in both cases, what these animals lack in productivity compared with their domesticated cousins, they more than make up for by commanding high prices from specialty meat markets, game preserves and investors. "We sell them mostly to people who just want something different, or who buy and sell these animals as an investment," notes Leroy.

'Customers Find Us'

The USDA doesn't keep statistics on exotic animal sales, but it's clear they're becoming a big business. "It's just starting to take off," says rancher Betty Kelso, president of the Exotic Wildlife Association in Kerrville, Texas.

Kelso reports growing interest among ranchers, who find it increasingly difficult to turn a profit on cattle and want

HIGH-VALUE CROPS AND ENTERPRISES

to respond to increased consumer demand for low-fat meat. "Many health-conscious food buyers are turning to game (animals)," she says. "It takes a lot less to feed these animals, and if done carefully, (exotics) can be grazed right along with cattle on the range. How far this meat market goes depends on how the public accepts it."

Judging from the demand for the Kaufmans' buffalo steaks, public acceptance is already high. Orders from individuals and bulk buyers are coming in faster than they can be filled. "We're about two buffaloes behind," Carolyn quips. "We really haven't had to find the customers. They find us."

The Kaufmans sell 30 to 40 buffalo each year for around $1.70 a pound, dressed. They also sell a few each year to muzzle-loading enthusiasts for $1 per pound liveweight. These modern-day mountain men march out into the Kaufmans' pasture, dispatch their new purchase with a trusty, old buffalo gun, and haul the carcass away to be turned into jerky, buffalo robes, moccasins, powder horns and the like. That saves the farmers the cost of processing and delivery.

Leroy says buffalo don't deserve their tough-to-handle reputation. "They have to grow up with you. And they're actually easier on fences, since they won't lean on them and tear them down trying to reach for grass on the other side the way cattle will," he observes.

But llamas are the couple's favorite and most profitable animals. "Last auction we went to, bred does were selling for $10,000 to $52,000. When we left the auction, more than $1 million worth had been sold — and they still had 90 left to go. And they're real healthy and just no trouble at all to raise," says Leroy, who keeps 14 llamas on a mere 2 acres of pasture.

In addition, llamas annually produce up to 10 pounds of wool, which sells for about $2 an *ounce*, compared with 50 to 60 cents a pound for sheep wool.

"Ostriches are real popular, too," says Leroy. Chicks sell for $500 to $800 each, while hens usually go for around $2,000. "I had an offer of $3,000 for one hen, but turned it down figuring I'd cover that with the sale of half a dozen chicks," he explains. Carolyn even uses the infertile eggs to make purses, lamps and music boxes that she sells at local craft shows.

The Kaufmans still grow about 125 acres of corn, oats and hay, which supplies most of the feed the animals require. Rabbit pellets and vegetable trimmings from local grocery stores are fed to their birds in the winter. "They sure eat a lot less than the hogs did," says Carolyn.

Like any "specialty" crop, exotic animals require different production and marketing skills from what most farmers are used to, say the Kaufmans. But to survive in farming today, even farmers producing more traditional commodities will have to change the way they do business. For the Kaufmans, switching to exotics was a natural — and profitable — way to accomplish that. As Leroy puts it, "We may have a menagerie, but it's also a profitable way of life."

Elk 'Rack Up Profits' In Many Ways

GAY MILLS, Wis. — "The United States imports more than 660,000 kilos (almost 1.5 million pounds) of venison from New Zealand. So there's definitely a market out there," says Robert S. Johnson, who raises about 75 elk on 518 acres of pasture and woods at Hardrock Game Farms in southwest Wisconsin. With prime cuts selling for as much as $10 a pound, elk are Johnson's most profitable enterprise. They're

Robert Johnson of Hardrock Game Farms grosses $1,000 from each elk sold for meat, and up to 10 times that from live trophy bulls sold to hunting preserves.

Photo courtesy of The Milwaukee Journal

even more profitable than the buffalo, bears, zebras, foxes, ferrets and other exotic animals he and his father raise with an additional 75 elk on a separate farm near Rockford, Ill.

But the Johnsons sell only 10 percent of their elk—mostly barren cows—for meat. That's because they can earn even more by selling live trophy bulls. "Hunting preserves need farm-raised elk, because wild elk will break their necks trying to crash through fences. And they'll pay $5,000 to $10,000 for them," explains Johnson. They also ship young bulls to the Orient, where the ground antlers are prized for aphrodisiacs.

Clever marketing is helping the Johnsons bring in quick cash to expand their operation. Investors buy a bull calf from them for $1,000 and pay a $1-a-day maintenance fee, "all of which the investor can write off on his taxes," notes Johnson. The customer gets the profits from annual antler sales (a 5-year-old bull's rack weighs up to 15 pounds and sells for $25 to $150 a pound), then cashes in when the animal reaches trophy size and is sold to a hunting preserve.

"We can make a person $10,000 in eight years on a $1,000 investment. It's a good deal for him and a good deal for us, since we can still keep the bull around to help increase our herd," says Johnson.

And except for the 8-foot-high fence required by Wisconsin law, elk aren't expensive to raise. "They're more like cattle than deer," reports Johnson, explaining that the elk's diet is normally 50 percent grass, while deer prefer twigs and brush. "But you can run a couple of elk on the same ground as a single head of cattle," he notes.

Johnson makes hay for winter feed from the same fields he uses as elk pastures in summer. "The elk will still paw through the snow for grass, even with hay available," he observes. Other than worming twice a year, health costs are minimal.

The market for elk meat is just getting organized, notes Johnson. He and his father sell prime cuts directly to restaurants for up to $10 a pound, while elk "hamburger" sells for about $4 a pound. They sell lower-quality meat to manufacturers of elk sticks and jerky. "We think a lot more jerky and elk sticks can be sold through specialty cheese shops and the like," says Johnson. Elk dress out between 200 and 300 pounds, and the Johnsons estimate they gross about $1,000 on each carcass sold for meat.

They also like the security of having many options for marketing their elk: meat, antlers, calves and trophy bulls. But there are risks getting started. "You should learn before you do it yourself," warns Johnson. "It's best to get involved with a breeder and learn the trade, unless you have money to blow."

Fallow Deer Net $700/A

MORAVIA, N.Y.—In 1985, Peter Duenkelsbuehler sold 300 fallow deer to farmers getting started in this unique livestock enterprise. Gross receipts from the sales: $90,000.

He butchered another 60 in his own slaughterhouse that winter. And even though the carcasses only dressed out between 40 and 60 pounds, Duenkelsbuehler grossed $18,000 by selling the meat for $6 a pound directly to gourmet restaurants and individual customers seeking lean meats.

"Fallow deer only have 44 milligrams (per 100g) of cholesterol, compared to 58 for chicken breast," says Duenkelsbuehler, explaining why meat from the small, spotted relative of white-tailed deer is becoming more and more popular in health-food markets.

Duenkelsbuehler runs the deer on pasture in summer, and feeds hay, oats and apple branches (for vitamins and minerals) in winter. He gets an even higher premium for his meat, because he uses no chemicals on his home-grown feeds.

To get started on 10 acres of pasture, Duenkelsbuehler estimates a farmer would need about $22,000. That would cover the cost of buying 40 does and four bucks, and building the 7-foot-high fence needed to keep the deer in (and dogs out). "There won't be any income until the first bucks are harvested 12 to 24 months later. But once you get going, you can expect to (net) about $7,000 a year," he says.

Peter Duenkelsbuehler grossed $18,000 from fallow deer he butchered in his own slaughterhouse, and $90,000 from live animals sold to farmers looking to cash in on this new market.

HIGH-VALUE CROPS AND ENTERPRISES

Fallow deer markets are well-established in Europe, where the animals originated. Duenkelsbuehler reports there are more than 2,000 producers in Germany, alone. And he's betting they'll catch on here, too.

$30 For 2 Pounds Of Pheasant

HOLLIDAYSBURG, Pa. — "Our dressed pheasants sell for about $3.50 a pound," says Dave Creuzberger, co-owner of the 132-acre Cross Keys Pheasantry. "That's pretty good when you consider chicken wholesales around 50 cents."

But it's peanuts compared with Cross Key's most profitable item: a 2-pound, smoked pheasant gift pack that Creuzberger will ship anywhere in the country for $29.95. Last year, he thought 1,000 would easily cover his orders, but he quickly sold out. "You talk about the cobbler's kids going barefoot, my mother-in-law didn't even get one," he jokes.

Cross Keys hatches about 100,000 birds each year. And even though Creuzberger sells 90 percent of them as chicks for 80 cents to $1.10 each, he brings in almost as much selling a relatively small number of full-grown birds. "We start selling live, full-grown birds from the spring hatching in September for $7," says Creuzberger. His main customers are game clubs and farmers trying to re-establish pheasant populations.

To cut feed costs, Creuzberger plants strips of forage sorghum in the pens to harbor insects and supply cover and vegetation for the pheasants. "People don't realize that pheasants eat lots of insects and vegetation. They only go for grain after frost kills all the insects." He also grows about 15 acres of corn to feed mature birds and overwinter his breeding flock.

Creuzberger's high prices are offset somewhat by higher costs and lower production. He only gets about 50 eggs a year from each hen in his 2,500-bird breeder flock. The eggs are gathered by hand from 20 acres of range pens built from 2-inch mesh netting strung 12 to 15 feet high. "I never stopped to figure out how much it would cost to build that many range pens from scratch, but it's not cheap," says Creuzberger.

Still, if a farmer just wants to make his land more attractive to hunters, raising a small pheasant flock is easy, says Holly Hollister of Cherry Bend Pheasant Farm in Sabina, Ohio. "All you need is a box and a light bulb to raise a few chicks. And (fee hunting) could become a good way to supplement farm income."

For More Information

The following groups can help you get started raising unusual animals. Some provide fact sheets for new owners, others publish newsletters and directories of breeders and sellers. Write for details.

Exotic Wildlife Association, 1811-A Junction Highway, Kerrville, Texas 78028. Phone: (512) 895-4288.

The American Minor Breeds Conservancy, Box 477, Pittsboro, N.C. 27312. Phone: (919) 542-5704.

Animal Finders Guide, C/O Pat Hoctor, P.O. Box 99, Prairie Creek, Ind. 47869. Phone: (812) 898-2701.

Llamas, P.O. Box 100, Herald, Calif. 95638

International Llama Association, P.O. Box 11530, Bainbridge Island, Wash. 98110. Phone: (206) 842-1614.

American Buffalo Journal, American Buffalo Association, P.O. Box 16660, Denver, Colo. 80216. Phone: (303) 292-2833.

Stagline, New Zealand Deer Farmers Association Inc., 4th Floor, Agriculture House, Johnston Street, P.O. Box 2678, Wellington, New Zealand.

North American Deer Farmers Association, C/O Josef Von Kerckerinck, Lucky Star Ranch, Box 273, Chaumont, N.Y. 13622. Phone: (315) 649-5519.

American Pheasant and Waterfowl Magazine, American Pheasant and Waterfowl Association, C/O Lloyd Ure, Rt. 1, Granton, Wis. 54436. Phone: (715) 238-7291.

Freezer Beef And Lamb From A Few Acres

This Tennessee family returned some worn-out land to productivity and is making it pay.

McMINNVILLE, Tenn.—When Ron and Pam Castle first saw the 12 acres of land they now own near here, they imagined what it would look like if it wasn't choked with ragweed. They tried to look past the thick clumps of sumac to see the walnuts, hickories and river birch, and wondered if they could heal the gullies that had carried away much of the topsoil. Four years later, the Castles have not only restored the farm, but have developed a reliable direct market for their freezer beef, lamb and produce, and are well on their way toward showing their first profit.

"One of our strong suits has been our willingness to keep our customers informed regarding quality meat, processing and feeding programs, and non-chemical methods of farming," says Ron. "They understand and appreciate the extra effort that goes into our product."

The Castles started out selling freezer beef. When their sheep flock came into production, they invited their 50 customers to their farm for a cookout—a lamb cookout. The guests didn't come empty-handed, though. They brought a range of tasty side dishes and famous Southern desserts to round out the feast.

Although busy telling fish stories, turning juicy, home-raised lamb chops on the charcoal grill, and checking the lamb roast in the hickory smoker, Ron never missed a chance to list the many fine qualities of lamb. Succulent lamb kebabs, chops, ribs, and smoked roast disappeared quickly, with much appreciative comment.

His visitors must have been impressed with the lamb, Ron says. He received six new orders, and some regular customers increased their orders.

While having enough room to entertain friends and grow healthy food for their four children is important to Ron and Pam, they're determined to make the farm a financial success, too. Their 10 acres of pasture/hay land provides much of the feed for finishing four to six head of beef per year and for raising 30 lambs, while the garden provides plenty of produce for the family and extra to sell at market.

Livestock From The Start

The Castles decided to get started in farming when Ron's job as an industrial implement salesman brought him to the rural McMinnville area in 1980. After four months of searching, they found their present farmstead, with a modern brick house in good shape, but with fields and barn in need of a lot of attention.

"The first thing we did was fence the perimeter of the land we knew we'd want for pasture," says Pam. They decided that, with all that land needing to be cleared of brush, why not get some goats to help out with the job? Soon, eight goats were turned loose in the fields and happily spent their days munching down the overgrowth. "You hear lots of stories about how difficult goats are to manage, but they worked really well for us," says Pam. "Only pregnant and nursing mothers needed any supplemental feed."

Bigger brush, logs, and other items the goats couldn't handle were removed by hand and placed in the gullies that had formed while the land was in row crops. After one season of work, the land was beginning to look like usable pasture.

The Castles also began upgrading the small barn that came with the property, building more fence, planting a small orchard in the backyard, and grooming a nice flat acre for a garden.

In 1981, the "clearing goats" were sold, and Ron overseeded the pastures with crimson clover and fescue, using a strap-on, hand-cranked, walking-type seeder that had belonged to his grandfather. Next, feeder cows were brought in.

"We decided to keep livestock from the start, because the land is suited to it," says Pam. "We wanted to raise enough for ourselves and enough extra to pay for it."

Sheep 'Appropriate'

Ron was also interested in sheep. When he had the chance to buy some registered Suffolks from an Illinois farm, Ron took home a ram and several ewes, with an eye toward providing good lambs for sale to other farmers and to his fast-growing freezer-meat market. His lambs sell well and his flocks make excellent showings at county and regional fairs, taking "best flock" and a number of other top honors.

"We're realizing that sheep are more appropriate for our small scale," says Ron. "We could carry up to 40 ewes here, making good use of our resources."

So in 1983, the Castles quit stocking beef cattle over the winter. In spring, they now buy 400- to 500-pound calves which will graze through the growing season, to be slaughtered and custom processed in September.

Ron Castle charcoal grills some lamb chops for his customer-guests who, in addition to having a good time, are learning how delicious well-prepared lamb can be.

Planned Marketing

"None of this would be possible if we were simply selling through the usual channels," says Ron. The Castles have worked out an integrated, skip-the-middlemen processing and marketing plan (see "Let Your Software Do The Selling" on page 124) that brings good prices and provides a good deal for customers. For example, in 1983, lamb sold at $1 per pound live weight, plus a processing fee of $20 per head. That put finished lamb into customers' freezers for $2.30 per pound, while supermarket prices ranged from $3.50 to $5 per pound. Beef is cut to order and sold by the quarter or half.

Equipment Needs Increasing

Their equipment inventory includes a 1947 model "M" John Deere tractor. "It's as old as I am," says Ron. Then there's a used moldboard plow, a mower, a new Planet Jr. planter, a used disk and a dump rake. The latter is something of an antique, but it's useful for what little haying Ron does. The tractor-drawn rake makes small stacks of hay that are then loaded onto a wagon and stored loose. Ron also buys 200 bales of hay to carry his livestock through the winter months when there's no grazing. As he brings the pastures into greater productivity, he would like to upgrade his machinery line and do more of his own haying. Ron also wants to increase the amount of permanent fencing within his woven-wire perimeter fence. He now creates four to six paddocks within the 10-acre pasture using electric "poly-wire" on portable reels.

Ron and Pam are resisting the urge to buy some adjoining land. "After the land fever subsides, we realize that we don't have time to do a decent job with more," Ron says. "We've decided to go for quality and maximum use of what we have."

The Kiwis Are Coming
They're bringing new profit potential to farms on the West Coast and in the Deep South.

CHICO, Calif. — Dick and Lisbeth Harter aren't ones to rest on a reputation. Since 1968, their 900-acre Cherokee Ranch in north-central California has quietly become one of the leading organic rice producers in the state — if not the entire country.

But in the mid-'70s, the couple decided to branch out — literally — into kiwifruit. By the early '80's, due to shrewd soil management and infectious promotional zeal, they gross nearly $30,000 a year from just 2 acres of this brownish, tennis ball-sized fruit that once could be found growing wild in the Yangtze River Valley of China.

"They looked like a fascinating new crop," says Lisbeth, who retired from nursing in 1973 and immediately became caretaker of the family's first 54 kiwi vines. "Dick loves a challenge . . . he didn't have time to nurse them along, so he said, 'This can be your project.'"

Apparently, Lisbeth made the transition from nursing to farming rather well. In 1983, the 290 kiwi plants on her 2-acre plot produced 50,000 pounds of fruit—a bumper crop by California standards—and grossed $28,000. That doesn't include the thousands of pounds left to ripen on the vine for later sale through a network of organic produce shippers. Dick says sales from this 'second' crop virtually paid for the hired help he needed during harvest.

Encouraged by that progress, the Harters soon added 3 acres to their kiwi planting. "They're going to do themselves proud again this year," Lisbeth says, estimating the 1987 harvest from all 5 acres at about 100,000 pounds. Although the kiwi market in 1987 was depressed, with prices averaging about 33 cents a pound, the, Harters were being offered nearly twice the going rate for their organic kiwi. "They are very rewarding, and they are fun to work with," Lisbeth adds.

Kiwi plants are dioecious, meaning male and female flowers are on different vines. They are planted in rows that, if viewed from overhead, look like a series of tic-tac-toe boards with a male plant occupying the center square, surrounded by eight female plants. Cordons (branches) are trained along from one to five trellises supported in wash-line fashion by crossarms.

Lisbeth Harter fastens mature kiwifruit to trellis. Cover crop of bell beans, oats and other legumes and grasses cools soil and adds organic matter.

This 2-year-old kiwifruit vine will begin bearing in two years, and will mature about four years later.

Since it's impossible to tell the sex of the plants from seed, kiwifruit are usually planted from 2-year-old cuttings that are about 6 feet tall. "Most people buy them already grafted from a nursery," Lisbeth explains. "From that point, it usually takes two years until you get a crop." Once vines start bearing, though, they waste little time making up for lost production. "Each female plant has the potential to produce around 400 pounds of fruit each year," says Lisbeth, noting that the vines triple their output annually until reaching the 400-pound peak.

In their first year, the Harters lost several plants and got just 200 pounds of fruit per acre, largely because the summer sun turned their bare, sandy soil into an outdoor griddle. They solved the problem by planting clover between the rows to lower soil temperatures and, of course, add nitrogen. "Then they really took off," Lisbeth recalls. Yields jumped to 800 pounds per acre the following year, and a whopping 16,000 pounds per acre the year after that.

Fall Cover Needed

As the first vines matured, their 5- to 8-inch-long, heart-shaped leaves shaded and killed the clover, forcing the Harters to find an alternate ground cover/green manure that would grow while the kiwis were dormant. Bell beans, a winter-hardy legume that can be sown in fall in California, were ideal. Dick says they add up to 100 pounds of nitrogen per acre to his soil, and have improved his organic matter level considerably.

Each month from February through August or September, the Harters also spread 1 to 2 tons of poultry manure per acre on their kiwifruit. "We just spread it," Lisbeth stresses. "We don't till the soil under the kiwis at all. Because we're on this sand, we've had to feed (and water) them more vigorously. The sand doesn't have any nutritional value."

The manure applications add 120 pounds of additional nitrogen per acre, but cost $28 each. That's one reason why production costs for the Harters' organic kiwis are a bit higher than those of conventional growers. It's also the main reason their yields are higher and more stable, says Dick. In 1983, he points out, heavy rains reduced California's kiwifruit production by nearly 25 percent, from about 4 million pounds to just a little more than 3 million pounds. "Our overall production was up about 34 percent, and our production of number one grades was up something like 84 percent. Where people are using chemical fertilizers, normally you don't have the earthworms and much organic matter in the soil," Dick adds.

Unlike most kiwifruit growers, who often train vines along just one wire in hope of getting larger fruit, the Harters string five wires. This should result in higher yields, but smaller fruits. Yet the couple say they get as high a percentage of large fruits as other producers do. "We get perhaps a larger crop. They'll break the crossarms if they're not supported correctly," says Lisbeth. "We get more large fruits than the average grower. The vines are so much healthier.... We've improved the organic matter in our sandy soil with the (cover crop) and the manure."

Weeds are controlled by mowing. The scales and leafrollers that plague many kiwifruit operations cause little damage to the Harters' crop. "The other bugs get them first," quips Lisbeth. "So far, we haven't run into any insects that bother them.

"Because we don't spray, we have a lot more pollination from the other bugs that crawl in and out," she adds, explaining that bees seldom visit the nectarless blossoms.

Healthier Than Citrus

Kiwifruit prefer a sandy loam soil, but can adapt to a wide range of soil types, from sandy to adobe. The Harters' newer plantings, in fact, are on an adobe soil and require drastically different management, Lisbeth says. Kiwis are grown mainly in California, with minor production in Oregon, Washington, Texas, Arkansas, Georgia and Florida.

Though mature vines can withstand frost, tender shoots and fruits are highly susceptible to frost damage. The fruits, which can be eaten fresh (peel the skin, first) or used in jellies, pie fillings and toppings, contain an enzyme that makes them popular as a meat tenderizer. They are said to contain more vitamin C, iron and other minerals than oranges. The growing season is about 210 days.

The crop is hand-picked in October and November, and immediately stored at 32 F and 99-percent humidity to prevent drying. According to fruit packers familiar with the Harters' crop, it's here that organic production methods really pay off. "They do store better. They're beautiful fruit," says Mark Kosterman, field operations manager for Alkop Farms. Alkop, which grows and packs kiwifruit exclusively and is California's second largest distributor of the fruit, has handled the Harters' crop for several years. The company is now trying to market a separate line of organic kiwifruit for a few other growers in the area, too.

"There's a slight advantage in that (the Harters) don't use chemical fertilizers," adds Kosterman. Their fruit has a slightly lower sugar content, which is better for storage. With conventionally grown kiwifruit, he says, excess nitrogen from highly soluble fertilizers creates more sugar and causes the fruit to ripen faster than normal. That's why kiwifruit growers around Chico cannot fertilize their crop after June 1, Kosterman explains.

"The organic fruit seems to hold really well—at least it looks that way," says Jim Callender, president of Callender Farms Packing Co., which first handled the Harters' kiwifruit in 1983. "Plus, there are no pesticides. There are people out there who are looking to build the organic market. We're hoping to pick up more (organic growers)."

How To Grow Kiwifruit

Kiwifruit, the exotic fruit from New Zealand, has been labeled by some as "the new fountain of youth." That nickname stems from the fact that members of a tribe in the interior of New Zealand who live almost exclusively on the oversized berry lead extraordinarily long lives, often up to 120 years!

Part of that may be due to the fact that kiwifruit is a very nutritious food. It has about twice the vitamin C of a medium orange, and contains twice the amount of vitamin E of an avocado, with only 60 percent of its calories. Kiwifruit is rich in folic acid, high in fiber, and packs more potassium per ounce than bananas. In all, kiwifruit contains some 20 nutrients, 17 of them in meaningful amounts. In that regard, it resembles a vegetable more than a fruit.

But some researchers believe that the kiwi's real secret is a little-known enzyme that, when consumed regularly, retards the aging process in humans. "That is why such extensive testing and study is going on with the tribe that uses it daily," says Franklin Hermsely, a scientist who is working with the tribe. "It's not only that this tribe's people are living such long lives, but they are in such healthy condition. Both men and women use the meat of the fruit as a facial treatment and have amazingly wrinkle-free complexions."

Acreage Doubles Each Year

That is also why kiwifruit is proving to be a profitable new cash crop for some farmers. Key markets are in Europe, Japan, Australia, the United States and Canada. Acreage planted to the fruit has doubled each year for the last 10 years. The price runs up to $1.60 a pound. "Kiwifruit is the new glamour crop. The prospect is for the farmer to gross $8,000 an acre," says James A. Beutel, an Extension pomologist with the University of California at Davis.

Originally known as the Chinese gooseberry, the fruit was renamed by New Zealanders after the Kiwi, their national bird. For successful kiwi production, you need a climate that provides a growing season of 225 to 240 days, 600 to 700 chilling hours with the temperature below 45 degrees, soil pH 6.0 to 7.2 and drip (or trickle) irrigation. A kiwi vine requires about 60 gallons of water a day. If you cannot provide irrigation, do not attempt to grow this crop! (A cold-hardy kiwi that survives freezing temperatures in the far north was introduced a few years ago. Its fruit resembles grapes more than the traditional, egg-sized kiwifruit with fuzzy brown skin that you see in groceries. But the taste is still that of a kiwi. One source of cold-hardy kiwi plants, a crop that I heartily recommend for northern farmers, is: Mellingers Inc., 2340 W. South Range St., North Lima, Ohio 44452.)

HIGH-VALUE CROPS AND ENTERPRISES

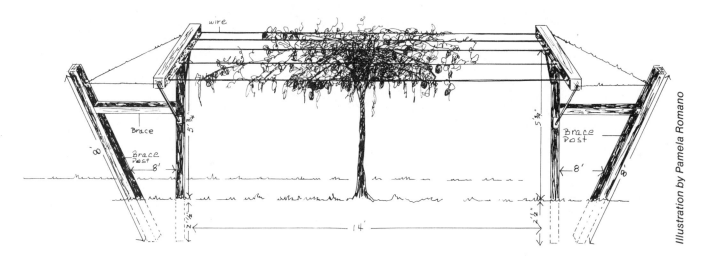

Illustration by Pamela Romano

You may establish your kiwi with seedling plants. I strongly recommend the use of rooted cuttings. The HAYWARD variety will begin to reach maximum production in its sixth year after planting, averaging 500 to 600 fruits per plant.

That kind of production should give you a clear idea of why you need to trellis kiwifruit and just how strong that trellising must be. I recommend the 5-wire, T-bar trellis system. All lumber should be pressure treated to resist decay. Use 8-foot posts. Line posts may be four inches in diameter, end posts six inches and brace posts 12 inches. Brace posts should be set eight feet away from end posts on each row and slanted away from end posts at a 30 degree angle. Set all posts 2½ feet in the ground. Attach a 6-foot crossbar to the top of each line and end post.

Number 9 crimped, galvanized wire should be used for support wires. Theoretically, crimped wire will not expand or contract and will maintain proper tension on the wires. Smooth No. 9 wire may also be used if crimped wire is not available. The center wire of the 5-wire trellis should be attached to the middle of the T-bar crossarms. The next two wires are placed on either side of the center wire at 18-inch intervals. Position the remaining two wires at the ends of the crossarms.

Once you have installed your trellising with the rows running north and south, it is time to plant. You may first want to guard against root knot nematodes by fumigating the soil with methyl bromide. Recommended spacing is 15 by 15 feet, for 193 plants per acre. The leading varieties are HAYWARD (female) and CHICO (male). One male plant is required for each 10 to 12 females. You want to dig a 24-inch hole, 30 inches deep, midway between each end post and the first line post. To each hole, add one bushel of moist peat moss. Set a plant in the center of each hole and fill the hole with soil. Prune plants back to just three buds above the graft union.

Then, tie one end of a length of nylon bailing twine to each plant and the other end to the center wire in the trellis. Select the strongest shoot and train it up the twine to the center wire to become the trunk. Remove all other growth on that shoot as it appears. When the shoot reaches the center wire, pinch out the growing point to induce lateral branching. Select two strong laterals and, using black plastic chain ties, train the laterals just below the center wire in opposite directions to the next line posts. When these laterals reach the line posts, the growing points should be removed. That terminates horizontal growth and stimulates lateral shoot growth. This now becomes the permanent arm or leader. Shoots from these leaders should be trained over the two wires on both sides of the center wire and held in place with black plastic chain ties. Once the shoots reach the outside wires, they should be trained or "combed" downward.

Kiwifruit must be protected from wind, which causes blemishes and makes the fruit unmarketable by rubbing fruit together on the vine. The HAYWARD cultivar is especially susceptible to wind rub. A windbreak should be 25 to 30 feet away from the vines and surround the planting on all sides. The ideal windbreak is fast-growing, leguminous, flowers the entire growing season and produces copious amounts of pollen and nectar. In the Southern United States, the species meeting all of those specifications is the Chinese tallow tree *(Sapium sebiferum)*.

Since kiwifruit produce only a small number of blossoms, proper pollination is the key to achieving the 90-percent fruit set required for a good commercial crop. Fruit size is correlated with seed numbers, which, in turn, depend on the degree of pollination. A good sized HAYWARD kiwi should contain 1,000 to 1,400 seeds, while a small, inadequately pollinated fruit may contain only 50 to 100 seeds.

'8 Girls For Every Boy'

The first essential for good pollination is the presence of male plants with a flowering period that matches that of the female cultivar being grown. All male cultivars produce viable pollen for the first two to three days after blossoming. Female flowers, on the other hand, are receptive to pollen for seven to nine days after blossoming.

It is important that the male plants are spread throughout the vineyard, with every female being no more than 25 feet from a male. This may be achieved by a 1-to-8 male-to-female planting ratio. To enhance pollination, I recommend that 10 strong hives of honeybees be placed in the center of each block of plants just before your plants flower.

The blocks should not be larger than 2.5 acres, each. Here is a planting diagram for achieving the right balance of male(M) and female(X) plants.

```
X X X X X X
X M X X M X
X X X X X X
X X X X X X
X M X X M X
X X X X X X
X X X X X X
X M X X M X
X X X X X X
```

Pruning

Pruning is one of the most essential aspects of vine management, and plays a major role in consistently obtaining good yields. This operation should be performed during the dormant period after the most severe part of winter has passed.

Successful pruning involves keeping the vine from becoming dense and tangled. So-called open pruning permits access for bees during flowering, and allows penetration of sprays. It also provides good air movement and light penetration to help minimize conditions that favor fungal diseases, such as *Botrytis*. Ample light also helps to ripen fruit and mature fruiting canes for the following season.

Only under conditions of adequate light penetration will new shoots originate from the desired points on or near the permanent arms of a vine. Shoots should be pruned back to 4-bud fruiting spurs. The objective is to leave the optimum amount of wood from the previous season evenly distributed on the vine. Fruiting spurs should originate on or near the permanent arm. Four-bud fruiting spurs should be five to seven inches apart and well exposed to sunlight. Normally, only the first six buds of a fruiting shoot are productive, since the kiwi bears fruit only on the current season's growth.

Each vine must be restricted to its allotted space and not allowed to tangle with its neighbor along the row or across the row. Trim the "hanging curtain" on the T-bar system to one foot above the ground. This trimming process should be continued all summer.

While the trunk is permanent, the leaders running along the trellis wires may, if necessary, be replaced by laying down a new shoot. Fruiting shoots may be renewed annually, or left for two or more years.

All that involves an awful lot of work, but the financial returns are well worth it. And there are other benefits, too. While it's probably way too late to do anything about my wrinkles, I do try to eat at least one kiwifruit every day. I like those little things. They taste like a combination of a strawberry and a banana. And, besides, maybe there is something to that business about living to be 120. Who knows?

Money Does Grow On These 'Trees'

It will be two years before your first harvest, but shiitake mushrooms can be a profitable new cash crop. They will help improve your woodland, too.

The interest in growing shiitake (she-e-ta-kay) mushrooms and in establishing new markets for this Oriental delicacy are clearly on the upswing, according to Joe Deden, director of the Forest Resource Center in Lanesboro, Minn.

Deden and other members of his staff have conducted seminars on producing shiitake mushrooms and developing markets in several Midwestern states. The seminars attracted about 4,000 people, mostly farmers looking for ideas to start sideline businesses.

The edible shiitake mushroom has great potential in the United States. The mushroom has high, economic value compared with the small-diameter hardwood logs on which it is grown. Shiitake is just now beginning to be commercially cultivated in the United States. But the single, largest factor limiting the growth of this new industry is lack of accessible information.

HIGH-VALUE CROPS AND ENTERPRISES

Hardwood plugs laced with shiitake mushroom spawn are placed in holes drilled in logs. Holes are then covered with paraffin to keep out "weed" fungi.

The techniques used to cultivate shiitake are not new. They have been successful in many countries, including the United States. However, new growers need to do preliminary testing to gain experience with shiitake. They will also have to adapt production methods to their particular climates or to different types of wood.

In Japan, more than 80 percent of shiitake is produced by small-scale farms. Being profitable, yet requiring low monetary investment, shiitake is often grown in family-operated businesses. Many growers use it as a seasonal cash crop to supplement their income.

A Nutritious Mushroom

Dry shiitake contains at least 20 percent protein by weight. This protein is higher in quality than most vegetable proteins, containing essential sulfur-containing amino acids that vegetable proteins often lack. The mushroom is also high in trace minerals and B vitamins, and low in fat and calories.

In Japan, shiitake mushrooms have been used as a folk medicine. If they are exposed to sunlight or other ultraviolet light sources, they can, like milk, become a good source of vitamin D. These mushrooms also appear to contain certain compounds which may have medical value. An interferon-inducing substance and factors that help the body reduce blood cholesterol levels have been extracted and are being studied.

Cultivating shiitake commercially involves:

- Obtaining seed culture, which is also called spawn or inoculum
- Finding and preparing suitable logs
- Inoculating the logs
- An incubation period of about two years, during which the shiitake fungus colonizes the logs
- Harvest, handling and marketing

Currently, spawn is obtained as live fungal cells actively growing on small pieces of doweling material or sawdust. Although more expensive, many growers say the plug-shaped dowels are more reliable and much easier to work with than sawdust. Hardwood plugs with tapered ends are very popular. Spawn usually comes in plastic bags or bottles.

The final quality and yield of this mushroom crop depend, in part, on the fungal strain that is used. They vary in:

- Preference for type of wood
- Resistance to weed fungi
- Time of first fruiting
- Ease of fruiting
- Ability to stimulate fruiting
- Temperature required for fruiting
- Season of fruiting
- General size, shape, intensity of color, and specific flavor characteristics
- Storage qualities

Oaks Work Best

Many kinds of U.S. hardwood species have been tested for shiitake cultivation. These include all oak species, chestnut, hornbeam, beech, alder, willow and maple. Some people have also had success with aspen and birch. Although they certainly need to be tested, many other hardwoods, such as hickory and elm, are expected to work. Tree species with thick, darkly colored bark seem to require the least effort for success.

Oak is perhaps the most reliable wood for cultivation. Small-diameter oak logs and branches seem to give consistently good mushroom production in the United States and Japan.

Wood that decays easily, such as aspen, birch and cottonwood, tends to give faster production, but it also tends to become contaminated more easily with weed fungi. Because small-diameter or bent trees can be used, the cultivation of shiitake can be coupled with weeding out undesirable trees in timber stands. Thus, the growing of this mushroom can be used as an effective method for timber management.

Once the logs are ready for inoculation, holes must be drilled for inoculation sites. Depending on log diameter, five to 30 inoculation holes are drilled in a symmetrically spaced pattern. If electric drills are used, only heavy-duty or industrial-grade, air-turbine driven drills will survive rigorous use.

After inoculation, the holes are sealed to keep the plugs from drying out and to stop the entry of weed fungi, which can be fierce competitors. Many kinds of non-toxic sealants, such as canning-grade paraffin, work very well. The important thing is that the sealant must not penetrate into the log and kill the shiitake fungus. If paraffin is used, a thin layer of hot wax is quickly brushed over the inoculation site and surrounding area, effectively sealing the inoculation site. Some growers also seal the ends of the logs and surface cuts where the wood is exposed with wax or a tree-pruning sealant.

Once inoculated, logs are usually left out-of-doors while the shiitake fungus incubates. If rainfall is insufficient and the logs become dehydrated, they should be watered thoroughly.

Let the log surface dry before watering again. During this incubation period, avoid frequent, light watering. Log surfaces that stay wet too long favor the growth of weed fungi. Discard logs that become badly contaminated with green or blue surface molds or that exclusively produce other mushrooms.

Optimal conditions for successful colonization of the log by shiitake are temperatures of 60 to 80 F, 80 percent to 85 percent relative humidity, and watering when necessary. You can tell if the fungus has been growing by looking for a white ring of fungal cells at the cut end of the logs. This should appear when the logs are moist and often within one year of inoculation. Cutting off the end of a representative log may help you see this more easily.

Usually, after two years of outdoor incubation, 4- to 6-inch logs are ready to produce mushrooms. Small logs may bear mushrooms somewhat earlier. The reason it takes so long for the first crop to appear is that the mushroom must first completely colonize the log.

To stimulate fruiting and maximize mushroom yields, logs ready to produce mushrooms can be soaked in water for at least six hours or a maximum of two days. The length of soaking depends solely on the time necessary to effectively soak the log. After soaking, the logs are then restacked to favor good mushroom production. Many growers prefer long, A-frame stacks. This stacking allows good ventilation, sufficient room for mushroom growth and easy harvest from both sides of the stacks.

In nature, this mushroom often fruits in the spring and fall after a rain when the temperatures are cool. Most strains of shiitake will not produce mushrooms out-of-doors during cold winters or dry summers. Once the mushrooms begin to grow, they often mature within three days to a week. At colder temperatures, mushrooms develop more slowly.

People who want to grow mushrooms year-round can construct lighted sheds to force scheduled production. Research on fruiting chambers is being conducted at the Forest Research Center in Lanesboro, Minn.

It takes about two years for inoculated logs to begin bearing. During that time, the logs must be constantly monitored and occasionally watered so that they don't completely dry out. Managing 500 to 1,000 logs is a good goal for a part-timer, while having 5,000 logs provides half-time income.

Harvest And Storage

The mushrooms are cut or snapped off at the log surface. They are then placed in ventilated containers and refrigerated.

Yearly mushroom yields depend, in part, on the diameter of the log and the number of years after inoculation. Small-diameter logs tend to bear mushrooms earlier and give somewhat higher overall yields than do larger logs—although slower, large logs tend to produce more continuously.

Actual mushroom yields vary greatly with experience of the grower. Experienced growers using 3- to 5-inch-diameter logs claim 20-percent to 30-percent yield—that is, 20 to 30 pounds total production of fresh mushrooms per 100 pounds of fresh logs over the entire life of the logs. A consistent 10-percent overall mushroom yield would be considered quite respectable.

Shiitake is a firm, dense mushroom. It doesn't bruise as easily as many other mushrooms. Its white gills also tend to resist discoloring. Fresh shiitake is a very popular and tasty product. When fresh and kept free of excess water, this mushroom has a shelf life of at least two to three times that of the common white mushroom. Also, if a market is currently unavailable, shiitake can be easily and quickly dried into a high-quality product with, of course, a greatly extended shelf life.

HIGH-VALUE CROPS AND ENTERPRISES

When shiitake mushrooms first start appearing, they're usually near inoculation sites. Once the mushroom spawn spreads throughout a log, mushrooms can pop up anywhere on a log.

Markets already exist in the United States for the sale of dried shiitake. Dried shiitake imported from the Orient is currently purchased by Oriental restaurants and import stores across North America.

The fresh shiitake mushrooms currently wholesale in the United States for $3 to $8 per pound. Growers marketing fresh, rather than dried shiitake stand much less chance of competition from growers in other states or countries.

Reprinted with permission from Summer 1986 issue of Rural Enterprises *magazine, P.O. Box 878, Menomonee Falls, Wis. 53051.*

For more information, contact:

Steve Bratkovich
Ohio Cooperative Extension Service
17 Standpipe Rd.
Jackson, Ohio 45640

Phone: (614) 286-2177

Joe Deden, Director
Forest Resource Center
Rt. 2 Box 156-A
Lanesboro, Minn. 55949

Phone: (507) 467-2437

Andy G. Hankins
Madison County Extension Agent
Box 10
Madison, Va. 22727

Phone: (703) 948-6881

Gary F. Leatham, Research Chemist
USDA—Forest Products Lab
P.O. Box 5130
Madison, Wis. 53705

Phone: (608) 264-5856

Editor's Note: *The big question for anyone thinking about raising shiitake mushrooms is not production, as much as cash flow: "How much money is it going to take, what will I get and when?" explains Dr. Paul Przybylowicz, laboratory director at Northwest Mycological Consultants, a company that manufactures and sells shiitake mushroom spawn (702 NW 4th, Corvallis, Ore. 97330. Phone: (503) 753-8198).*

Przybylowicz, who is writing a book on shiitake production, says inoculation costs $1.50 to $3 per log (includes spawn, labor and cost of acquiring log). Facilities and management will add another 30 cents to 60 cents per log. Total yield is 2.5 to 3 pounds per log over the three to four years a log is in production.

Fresh, ungraded shiitake were wholesaling for about $5.50 per pound in mid-1987. Retail prices were $10 to $12 per pound. Oriental growers produce dried shiitake so cheaply that drying "is not worth doing unless you have fresh shiitakes that are going to spoil," says Przybylowicz.

The best books currently available on shiitake production are:

- *"Growing Shiitake Commercially," by Bob Harris, Science Tech Publishers, Madison, Wis. 53705*
- *"How To Grow Forest Mushrooms (Shiitake)," by Daniel and Mau Kuo, Mushroom Technology Corp., Naperville, Ill. 60566*

And Now, The Profitable Pecan

I have not considered the pecan as a high-value crop in the past because it did not meet my requirement of grossing a minimum of $3,000 per acre per year. That's because the old, standard varieties such as STEWART, DESIRABLE and MAHAN require a spacing of 50 by 50 feet or 60 by 60 feet, which allowed planting of only about 12 trees per acre. With those varieties, if a grower's production averaged 200 pounds of nuts per tree per year and the nuts sold for 75 cents a pound, the gross annual return per acre would only be $1,800, which is well below the $3,000 minimum.

That is being generous, because the average pecan yield in Alabama is only 900 pounds—per acre. The reason the average yield is so low is because most farmers pay very little, if any, attention to their pecan trees. They'll do all kinds of things for their corn and soybeans, apply lime and fertilizer, cultivate, spray herbicides and insecticides, the whole bit. But they just won't pay attention to their pecan trees the way they should. They just let their trees take care of themselves. And anything that falls off of those trees is just manna from heaven, as far as they're concerned.

If farmers paid half as much attention to their pecan trees as they do their corn and soybeans, why they could push the average yield way up there. For proof of that, all you have to do is look at the three, 60-year-old pecan trees in my backyard. I harvest 1,100 to 1,200 pounds of nuts from just those three trees most years. How? By doing everything right, just the way I recommend small farmers handle every crop on their farm. I keep the soil at the right pH, fertilize and make sure those trees have plenty of water. Ordinarily, a pecan bears nuts in alternate years. But when you take care of it the way I just described, it will bear annually.

Apply those state-of-the-art management techniques to some of the new semidwarf pecan varieties like CHEYENNE, and watch out! You can plant about 70 of those trees per acre. After five or six years, they'll produce 20 to 30 pounds of nuts per tree. By the eighth year, they will increase to 40 to 50 pounds per tree per year.

Rent A Tree

Now that, in itself, is not enough to meet my $3,000 minimum, either. After all, an average of 25 pounds per tree returns only $1,312.50 from 70 trees, at 75 cents per pound. And even an average of 45 pounds per tree adds up to just $2,362 per acre. That still doesn't meet the $3,000 minimum. But once those trees start bearing, you should lease them to members of your Clientele Membership Club for $50 a year. By doing that, you can increase the per acre return to from $3,412 to $4,462.50!

You can do the same thing with English walnuts, pistachios and other kinds of nut trees. All it takes is good management and even better marketing. Just as soon as you lease a tree, you want to make up an attractive, little sign with the customer's name on it and put it right on that tree—because that's their tree, now. Their very own personal nut tree. You should encourage every person who rents a tree to come out to your farm and check on their tree as regularly as possible, so that they firmly believe that. Tell them they should bring the whole family, their relatives and friends to see this magnificent tree that will provide them with an abundance of delicious pecans in the fall. Before you know it, that tree could become the background for family portraits and even the subject of science reports for school.

Of course, you will have to take care of these trees throughout the year. But when it comes time to harvest, all you'll have to do is drive your tractor out into the orchard and shake the trees with your mechanical tree shaker. The nuts will come tumbling to the ground and your customers will scoop them right up. To make this easier for everyone—and eliminate any possible arguments over which nuts belong to which tree—it is a good idea to mark off the boundaries of each customer's tree by running 3-inch plastic ribbon midway between the trees. Be sure to weigh each customer's harvest, so you can keep track of each tree's production.

Two nurseries that are good sources of pecan trees are:

Bass Pecan Co.
P.O. Box 42
Lumberton, Miss. 39455
(601) 796-2461

Monticello Nursery
Monticello, Fla. 32344
(904) 997-3593

Your investment in pecan trees now will pay handsome dividends for many, many years to come. For example, the pecan trees that George Washington planted on his farm at Mount Vernon before the Revolutionary War are still producing nuts. Besides being long-lived, the pecan is an extremely complex and somewhat mysterious plant that can benefit you in unusual ways. Why just watching a pecan tree's growth in spring can help you manage your other crops. There is an old saying in the South that when the pecan tree leafs out and the leaves reach the size of a squirrel's ear, it is then safe to transplant tender crops such as tomatoes and peppers. That's all I used to hear from my granddaddy when I was growing up. I thought he was just being superstitious. Then I went off to college and learned that you could just about bank on that folk wisdom. For some reason, no one has figured it out yet, pecans are almost never caught by late frosts, maybe two or three times in the last 100 years. How many of your other crops are as smart as that?

Why Rosemary And Basil Ride The Subway

BY TERI AGINS
Staff Reporter for The Wall Street Journal

NEW YORK—When New York's fanciest chefs need the freshest herbs, they often get them right off a farm in the South—the South Bronx, that is: an urban eyesore of barren lots and mean streets, shadowed by the hulks of half-demolished buildings in one of the country's most depressed communities.

An innovative agricultural enterprise, Glie Farms, Inc., with some venture capital from W. Averell Harriman Holding Company, has solidly and profitably rooted itself in the local landscape. On a block it shares with a small primary school and three storefront churches (two Pentecostal and one Jehovah's Witness), Glie Farms runs a thriving commercial greenhouse and herb nursery. Before Glie (pronounced Glee), about the only herb of much commercial importance in the neighborhood was one that went into funny cigarettes and wasn't oregano.

A cloying fragrance fills the rickety two-story house that contains the Glie packing plant. Women on an assembly line chatter in Spanish as they spank the dirt from handfuls of tarragon. Cut and weighed, the herbs are sealed in plastic bags and styrofoam cartons that will be shipped to such places as Le Cirque, the Grand Hyatt Hotel and Bloomingdale's Le Train Bleu restaurant.

Speedy Delivery

On a typical day, Glie ships 20 pounds of basil, 100 pounds of tarragon and 50 pounds of rosemary. Sometimes, in the interest of speed, employees deliver the goods by way of the New York subway system.

Glie sells 32 different herbs, including rare ones like purple basil and lemon verbena (used in place of lemon rind), to more than 200 of New York's toniest restaurants and hotels, as well as to supermarkets. "I applaud them," Chef Andre Joanlanna of La Grenouille says of the Glie staff. He orders his tarragon and edible flowers, such as nasturtiums, once a week from Glie.

Gary Waldron, foreground, and Glie Farms employees proudly display their aromatic, but profitable produce.

Photo by Bob Kinmonth

Glie is now selling $80,000 in herbs per month. At that rate, its 1985 sales will reach about $1 million, more than double its 1984 volume. Glie won't disclose profits. But the business has operated successfully enough to pay its founder and president, 42-year-old Gary Waldron, a salary of $40,000 a year, up from the $20,000 he was paid when he started the company in 1981.

It all grew out of a non-profit employment program for runaway youngsters called the Group Live-In Experience (GLIE) in the South Bronx. In 1981, Mr. Waldron, then working as controller at an International Business Machines Corp. plant in Brooklyn, took a year's leave of absence to help develop the program with a $100,000 federal grant. Working with 15 troubled youngsters, he set them to growing mushrooms and later, herbs. After a year, he returned to IBM in a new $60,000-a-year job.

A Speedy Return

But Mr. Waldron, who had grown up in the South Bronx, found that his experience with GLIE had dulled his interest in conventional corporate life. After six days, he left IBM to return to the South Bronx project and turn it into a profit-making business.

Glie still owns a minority interest in the firm. Other minority holders are Mr. Waldron, the firm's 30 other employees, and a Dutch contractor who has just built a state-of-the-art hydroponic greenhouse for the firm in the nearby Bathgate Industrial Park, financed by the Port Authority of New York and New Jersey to help stimulate employment in the South Bronx. A controlling 51 percent interest is owned by Tair Ltd., a Houston venture-capital firm, and by a subsidiary of Merchant Sterling Corp., the Harriman Holding Company. "It's a type of project that public-minded people like Averell dream about," says William Rich, the vice president of Merchant Sterling.

Twenty-eight of Glie's 32 employees are black or Hispanic residents of the neighborhood who earn $5 to $5.25 an hour. The other four include Michael Dowgert, 27, who recently received a doctorate in plant physiology from Cornell University.

One of Glie's original employees is Syvilla Young, 55, who works in the greenhouse, harvesting nasturtiums, pinching their yellow and tangerine blossoms at the neck and grouping them into stubby bouquets. "Just look at all of them," she says. "They're really too pretty to eat." But the artsy Chanterelle Restaurant spends $40 or $50 per week on nasturtiums to garnish duck foie gras and salads. "The flowers are amusing, nice little things to play around with," says David Waltuck, Chanterelle's owner. But, he says, "Our waiters have to tell customers that they're edible."

President Waldron has ambitious plans for Glie. The firm is preparing to start a 200-acre farm in Puerto Rico to grow outdoor herbs year-round. It now supplies open-air herbs as a broker for California growers, but acting as a broker is less profitable than producing and selling its own output.

Meanwhile, Glie has just begun to harvest its first crops grown hydroponically—without soil—in its new greenhouse. Each plant sits in a block of rock wool soaked with liquid nutrients. Light, air, acidity, humidity and temperature are computer-controlled.

"This is probably the most advanced system in the country and certainly a first in herb growing," says Harry W. Janes, an associate professor of horticulture at Rutgers University in New Jersey. Mr. Janes monitors the greenhouse project for the Port Authority, which has helped finance it.

Mr. Waldron has become an exceptionally successful salesman. He recently obtained for Glie a $150,000 contract to undertake landscaping for Chemical Bank in New York. As his successes have mounted, so has his reputation as a businessman. A group of investors interested in expanding cable television to poor households has retained him as a consultant. "Once people see you can be successful in one area," he says, "they come to you for other solutions."

(Reprinted, by permission, from *The Wall Street Journal*, March 10, 1985)

Editor's Note: *By 1987, Glie Farms' work force had increased to 140 and annual sales hit $2.5 million. It now sells fresh herbs, flowers and specialty crops to more than 800 restaurants and markets nationwide. Gary Waldron attributes such success to "finding a market niche" and "having a superior product, superior service, and persistence."*

While similar opportunities await in all of our large cities around the country, farmers can also take advantage of this trend by planting small plots of popular herbs and flowers for members of their Clientele Membership Clubs.

Nature's 'Hydroponic' Harvest

This farmer found a profitable new cash crop right in his own backyard.

DAVE McCOY

FREDERICKTOWN, Ohio — I don't know how long watercress has been growing wild in the spring just behind our house. A safe guess would be a century or more, since this farm has been in our family since 1857 and is a registered Ohio Historical Homestead.

But I do know that more and more local chefs are using the peppery-tasting leafy green to spice-up salads, soups, sauces and stuffings. And because of that, I was able to earn about $120 selling the cuttings from just a few hundred square feet of watercress in 1986 and am hoping to increase that to $3,000 or $4,000 a year, soon.

Watercress, a member of the mustard family, grows naturally in fresh, moving water and along the banks of streams and lakes. On our farm, it grows alongside some of the pastures in a spring that supplies us with water. It was my uncle, a Naval officer in the Pacific during World War II, who first suggested we harvest the leafy perennial and sell it like the Japanese do. When I began farming the home place in 1979, I gave the idea some thought — and that's as far as it went.

But the more I learned about farming, the more I knew it was important to use *all* my resources, not just my field crops and cows. Then I read a magazine article about a Tennessee family that grows watercress. That gave me the boost I needed, and in the winter of 1985-'86, I began looking for local markets.

The search was frustrating for awhile. All of the logical buyers — restaurants and produce wholesalers — *said* they wanted "fresh" watercress. But they were buying the crop from Florida and California. (One wholesaler's motto is "The Fresh Approach." It is one approach to freshness, alright. His watercress sometimes sits on ice for weeks before it's delivered to his customers.)

I finally met a local specialty-produce supplier who showed some interest in my crop. Even he was reluctant, at first. But his attitude changed completely when I showed him my "certified, organic" certificate from the Ohio Ecological Food and Farm Association. Not only did he want my organic watercress, but he insisted I specify on each case that the crop was grown without chemicals.

The first week of May '86, I grabbed a sharp knife and started cutting the roughly foot-tall plants at water level. I knelt on plywood atop a ladder placed lengthwise across the plot, so that I wouldn't disturb the delicate roots or muddy the water. The whole family helped package the crop. We used rubber bands to tie the plants into bunches, each the size of a small bouquet of flowers.

We packed 24 bunches into a case, and sold the first six or seven cases for $6 apiece. (Commercially grown watercress wholesales for $4 to $12 a case, with $8 or $9 being about average.) In all, we sold about 15 or 20 cases to that one wholesaler in '86.

Homemade "solar" A-frames help extend Dave McCoy's watercress harvest into fall and winter.

Photos by Dave McCoy

Meet Year-Round Demand

One of the first things I learned when I took this little venture into the alternative-crop market is that food buyers demand a guaranteed, consistent supply. No wholesalers or grocery stores would buy my watercress just in spring and summer, when they could pick up the phone and have Florida or California deliver the crop year-round.

So, to be successful, I realized I'd have to find a way to extend my harvest. It seemed logical that watercress would grow throughout winter in my spring, since the water temperature remains a fairly constant 56 F. The trick was harnessing the sun's warmth to protect the plants in fall and winter.

My solution was to build small "solar" A-frames that could be removed or relocated easily. I got the idea from an Amish neighbor who built a permanent structure over his spring to allow him to harvest watercress in December. Since my watercress is located in a muck that won't support a heavy, permanent frame, I considered building a lightweight, portable greenhouse out of aluminum angle-iron or tubing from a junkyard. But I couldn't figure out how to attach the greenhouse plastic to the metal.

My Amish neighbor suggested 1½-inch plastic water pipe, and that turned out to be perfect. The pipe comes in 20-foot lengths, but it's too flexible to build the frames 20 feet long. Instead, I decided to make the units 10 feet square and about 3 feet tall. I built two of them, each with a support (made from the same pipe) in the middle to fortify them when snow falls.

I inserted plastic, 1½-inch plugs into the pipes at the base of each frame. The plugs protrude just enough to let me anchor the structure into the muck with U-shaped, ½- by 36-inch reinforcement bars (like the kind used in concrete).

I stretched used, 6 mil greenhouse plastic over the frames, by first rolling a 5-foot section onto a piece of wood lath, then attaching the lath with screws into the bottom of the frame. The plastic should last three years, provided no sharp objects fly into it. To keep drafts out, I shoveled the muck away from the base of the A-frame, forming a seal between the pipe and the soft ground.

Experiment Pays Off

We began experimenting with cold-weather harvesting early in 1987, and managed to cut five cases of watercress in February. A mild winter may have helped. We did learn that it will probably take two months or so for watercress to recover from cutting in fall and winter, compared with a month or less in spring and summer. Yields will likely be lower in winter, too, because the growth is reduced along the edge of the frame, where temperatures are lowest.

We also learned that our milk cows like watercress as much as gourmet consumers. So we've had to erect some high-tensile fencing along the border of the pasture.

To avoid overexposing the watercress to harsh winter weather, I plan to lift each A-frame off the area it covers, and quickly cut the plants before replacing the frame. I'll have a

McCoy checks foot-high watercress in February '87, just before harvesting. This plot was covered with an A-frame, but the dormant, darker-colored watercress behind him was not.

helper on each side to replace the frame after I'm finished cutting, because the frames are awkward for one person to handle, especially if it's windy.

Summer-storage is the only problem I still haven't solved. I currently store the frames outside, one stacked on top of the other. But ideally, they should be stored inside, because sunlight causes the plastic to deteriorate.

Each of the two A-frames took about three hours to build, and cost roughly $60 apiece. Of course, the only thing I had to buy was lath and pipe. I plan to build as many as 18 more, soon. A 30- by 30-foot length of greenhouse plastic (enough for two, 10- by 10-foot frames) costs $60, so my cost to build the additional units will be around $85 apiece.

Why consider building more A-frames? One reason is that I intend to keep looking for more places to sell the crop. But even more importantly, my buyer liked the first year's watercress crop so much that he told me he'd now take five times as much.

Based on our limited experience, it appears we'll be able to meet his demand. We expect to get six or seven cuttings a year, and average from five to eight cases per 100 square feet in each cutting. That means we should gross from $3,000 to $4,000 a year when we get into full production, based on the several hundred square feet of watercress we have.

We plan for this to be a family project so we can teach our three children (ages 2, 5 and 7) the value of work and the importance of free enterprise. The money earned from the watercress will be saved for their futures.

And although I've used the frames just for watercress, I suspect they could be used with most any 'aquatic' crop. Alfalfa or bean sprouts are two examples. With a little imagination, I'm sure you can think of others.

Dave McCoy milks 45 cows on his family's 205-acre farm near Fredericktown, Ohio.

HIGH-VALUE CROPS AND ENTERPRISES

Growing Watercress Commercially

Watercress is a hardy perennial that commonly grows in flowing water. It is high in calcium and vitamins C and A, and contains moderate amounts of several other elements. Commercial production probably began in Europe during the 18th and 19th century, and has since spread throughout the world.

In England, watercress grows best in chalk or limestone districts fed by deep springs or artesian wells whose water temperature remains about 50 F year-round. In the United States, it is found from the Southeast to Canada. Commercial production is highest in the Mid-Atlantic states, as well as California.

Cultivated watercress usually is grown in side-by-side beds, with water supplied by a feeder stream along the higher ends of the beds. Each bed should have its own inlet and outlet, and be separated from other beds by raised walkways.

Most commercial plantings are direct-seeded, or established in seedbeds and then transplanted. The beds should be level, cultivated lightly and slightly moist, with no water flowing at the time of planting. Seeding rates average 1 pound per 700 square yards.

The recommended planting time is late spring or early summer, which provides a first crop in early fall. Subsequent crops can be established from cuttings during June and July, and from pulled plants later in the season.

Thick stands (6-inch spacing on all sides of plants) yield best and help prevent weed growth. Transplants should be set in shallow, gently flowing water, which will generally provide all of the required nutrients except phosphorus. When newly seeded plants are about 1 inch tall, the amount of water in the beds may be increased gradually as the plants grow.

Watercress is harvested two different ways. From March to October, when the air is warm and the crop grows above the water, producers cut the plants with a sharp knife, leaving stubble below the water to regrow. When the air gets colder, the crop grows best underwater. During this period, watercress is pulled (actually, thinned), leaving as much as two-thirds of the stand intact. Pulled plants are gathered into handfuls and the roots are cut off.

After harvest, watercress is washed and tied into bunches. It is highly perishable, so it is often sprayed with cold water before shipment. Dark, green color is essential for market appeal; chlorotic (yellow) plants suggest manganese deficiency and excess zinc.

Adapted from "Leafy Salad Vegetables," by Edward Ryder (1979; AVI Publishing, Westport, Conn.)

Editor's Note: *Watercress seed is available from Johnny's Selected Seeds (411 Foss Hill Road, Albion, Maine 04910), Kester's (P.O. Box V, Omro, Wis. 54963), Burpee's (300 Park Avenue, Warminster, Pa. 18974) and Stokes (P.O. Box 548, Buffalo, N.Y. 14240-0548).*

Triple Your Tree Profits
These simple management and marketing tips can show you how.

Burnell Fischer picks up the telephone receiver. On the other end is a farmer sitting next to a timber buyer who's just offered the farmer $5,000 for some trees in his woodlot. "The farmer thinks the money might get him out of a jam. He wants to know if $5,000 is a good price," says Fischer, an Extension forestry specialist at Purdue University. "He'll say to me, 'Boy, I didn't think those trees were worth that much!' But if he spends just a little time on marketing, he can get more than $5,000—maybe two or three times more."

Fischer's tale isn't an isolated case. Fact is, professional foresters all over the country say farmers often lose thousands of dollars by underestimating the value of their woodlots and jumping at the first offer from a sawmill or other wood merchant.

Part of the problem is that woodlands simply aren't an integral part of many farms nowadays. "They're not used as a source of fence posts, as the source of fuel for the cookstove or as the main source of lumber," notes Ohio Extension Forester Steve Bratkovich. For this reason, many farmers make little or no attempt to improve the quality or productivity of their trees.

A lack of marketing skills is equally to blame. "Farmers are good at marketing corn and beans, but they may only sell trees once every 10 or 20 years without knowing what they're worth," says Fischer.

The results can be disastrous. "I just heard one about a farmer vacationing in Florida who got a call from a buyer offering $1,100 for stand of trees back home," recalls Bratkovich. "He said he didn't need $1,100 just then, and turned it down. Then another buyer—from the same company—calls the farmer's son, who's at the farm, and offers him $20,000. Now, that first buyer was not necessarily dishonest. Maybe he didn't know what he was doing. Maybe he just saw railroad ties in those trees, while the second buyer saw furniture. But here's a good example where different people will make different offers for the same stand."

Know Your Product

How can you avoid these common pitfalls? One way is to plan ahead, so you won't be caught unprepared when a buyer shows up with a sales agreement in hand. Loggers tend to solicit farmers "because they happen to be the owners of timberland," says forestry consultant Dwight Stewart of Manning, S.C. If you're approached, "It may be the time to harvest some trees. The market value for a certain species you have might be peaking, and the logger needs what you have."

To ensure that you cash in on that demand, Purdue's Fischer suggests having state or federal forestry experts visit your woodlot and determine its market potential. "Every state has a department of forestry that employs service foresters," he says. At no cost, public foresters will survey a woodlot for "valuable tree species."

"The forester can at least tell you whether you've got enough of something to sell," says Bratkovich. Public foresters can also suggest initial management techniques (thinning, for example) you can do yourself to improve the woodlot. In some states, they may even be able to help you sell the trees for a fair price.

And when you get their advice, follow it. Clell Solomon, supervisor of the Missouri Department of Forestry's Conservation Management Program, tells of one farmer with maturing walnut trees who "decided he'd save a little money and cut the trees himself. Sawmills like logs 12 feet long and up. But he cut them the wrong size, about 6 or 8 feet. Then he used the tractor to drag the logs across a plowed field, got them all mud-encrusted. They didn't look good at all. He was able to sell them, but for only a fraction of what they should have been worth."

Don't jump at the first offer that comes along. Timber prices vary widely.

Farmers with large, valuable tracts of maturing trees may need more extensive guidance than a public forester can provide. Here's where private consultants enter the picture. "They can do a lot of good," says Fischer. "They know who's looking for what, and can work for you as the middleman." Since the typical private consultant works for a percentage of the sale, "You know he'll try to get top dollar, because he stands to gain, too." State forest departments usually maintain listings of qualified consultants.

Stewart, who's arranged many such sales, agrees that a farmer could easily double his income from a stand if he calls on an expert for help. The very presence of a professional consultant is often enough to scare off low bidders who won't compete with the fair market value of wood, he says.

But buyers are businessman, Stewart warns. "If I were a bidder, I'd start out on the low side, too."

Plan For Profits

Assessing the value of your trees is only one component of the pre-harvest planning process, foresters stress. Planning what to do after the harvest is just as important.

"We try to encourage the farmer to think about reforestation before the sale," says George Kessler, Extension forest specialist at Clemson University. "For example, if he or the logger takes all his good trees, his options for (natural reseeding and regrowth) are gone."

B.D. Ashbaugh of Ava, Mo., learned that lesson the hard way. In the mid '50s, he purchased 2,800 acres of native white oak and black walnut from a timber company as an investment. He conceded two years' harvesting rights to the company. "When they pulled out, they left me with blackjack (oak), which nobody wants, and culls. We had to kill everything" and start over, he says.

Ashbaugh, now retired from a machine tool company he managed, has since developed a management plan that illustrates both the short- and long-term profit potential for his woodlot. The plan begins with 7,225 trees per acre planted on 2.5-foot centers that will naturally self-thin.

- At 15 years, the remaining 1,849 trees will be 3 inches in diameter. Pick out the best 1,387 trees and cut the rest, producing 7.39 cords of wood. Ashbaugh can sell the wood to local homeowners for $60 a cord. *Gross: $443 per acre.*

- At 25 years, trees are 6 inches in diameter. Remove 341 trees to produce 12.25 cords. *Gross: $735 per acre.*

- At 30 years, trees are 10 inches thick. Remove 60 for 5.87 cords. *Gross: $352 per acre.*

- At 45 years, trees are 14 inches thick. Remove 25, producing 1,975 board feet of timber per acre selling for $60 per 1,000 board feet. *Gross: $118 per acre.*

- At 60 years, all remaining mature trees, now spaced 40 feet apart, are 18 inches in diameter, producing 8,400 board feet of timber. *Gross: $507 per acre.*

Deciding on a method of payment before the sale contract is signed and the trees are cut is critical, too. "An extended pay period over more than one year may help shelter the timber sale income from capital gains taxes," says Clemson's Kessler.

A farmer need not cut his immature trees to gain some fast capital, either. There are investors out there who are willing to pay a premium price for ground with a good stand of quality trees, particularly if they are already accessible by a road, says Purdue's Fischer. Selling the acreage instead of just the trees may free a busy or distressed farmer from post-harvest replanting and thinning. "Time and money may be the last thing the farmer may have. If he waits 15 years (to manage and then harvest the maturing trees), he might not be in business anymore."

Lloyd Grafton, who grows cash-grains on 2,500 acres near Laclede, Mo., is in no hurry to harvest his 700 acres of native white oak and black walnut. "Harvest is a long ways off," he says. "Most farmers around here will tell you their acreage is 75 percent tillable and 25 percent wasteland. We don't have 25 percent wasteland. We have trees! We harvested about $1,400 worth of wood about six years ago, and we sell a couple thousand dollars worth of firewood a year, so we're not really profiting from it yet. But in time, we'll have a tremendous harvest."

Meantime, Grafton uses his thinnings to fire his modified grain driers, for a $3,000- to $4,000-per-year savings in natural gas.

When you think about it, profiting from a woodlot requires the same kind of careful thinking as any other enterprise, notes Vermont Extension Forester Thom McEvoy. "Most American businessmen make their money simply by knowing more about a product than someone else—be it used cars or timber. If a landowner later finds out that he could have gotten much more for his timber, he wasn't cheated. He just wasn't smart about the transaction."

Sell Trees On Your Terms

While you may not make a fortune from your woodlands, you can avoid losing a fortune by knowing the value of your trees before you're approached by a buyer. If you do receive an offer, beware of these common sales terms and pitfalls:

Diameter-limit cutting—gives loggers free rein to cut any tree over a specified diameter. Rather than cut all the trees that size, some loggers cut only the best trees, leaving behind the lower-value ones, which often aren't even fit for firewood. Avoid such an agreement.

Sloppy logging practices—Leftover debris or poorly laid logging roads can damage remaining high-value trees, reducing their vigor and reseeding value. Before signing a sales agreement, check with landowners who have previously dealt with the logger. Inspect sites frequently during harvest.

Uncontrolled harvests—can lead to the logger cutting trees you wish to preserve, or your neighbor's trees. Mark trees you don't want cut, and designate borders carefully.

Confusion over who's who—can lead you to view the logger as your forest manager. Worse yet, the buyer may mistake a tenant or a neighbor as the landowner. Spell out all relationships in writing.

"Henpenning"—an old horse-trading tactic, starts with an initial offer so low that most any landowner would reject it. Shortly afterward, second and third buyers (in association with the first) make slightly higher offers. By this time, a landowner, comparing the third offer with the first, thinks he's getting a good deal. In reality, the wood may be worth much more. Have a public or private forester determine the value of your woodlot ahead of time.

Mail-order sales—involve a landowner (usually absentee) who receives a price through the mail. The envelope includes a contract that the seller is asked to sign and return by mail. Before proceeding, review the contract with a forestry expert familiar with your stand.

Finding Those Hidden Dollars In Your Woodlot

There's more value in woodland than first meets the eye. Here are some tips on how to spot it.

GENE LOGSDON

Farmers who love their woodland with a passion (and it requires passion nowadays to save trees from the thirst for more grain land) like to tell the story of how an Indiana character known as Tain't So Brown justified his woodlot in the face of seemingly more lucrative corn and soybeans. "That 20-acre woods of yours sets right in the middle of land that can produce 150 bushel-per-acre corn," his cash grain neighbor told Brown, shaking his head sadly. "You got to bulldoze those trees out of there and start making that land pay." Tain't So shot the neighbor a withering look. "Tain't so, young man," he replied in his high, whiny voice. "Tain't so at all. Last year, when coonskins were bringing $30 apiece, that woods netted me nearly $50 an acre, not counting the pleasure of the hunting." The neighbor, taken aback because he had cleared only a little more than that on his corn, finally found his voice. "But you won't catch that many coons every year at that price," he replied. "And," shot back Tain't So, "you won't grow 150-bushel corn at a good price every year, either."

The smart money boys have all the figures to show that woodland can't compete with row crops in profits or cash flow. But, as Tain't So Brown points out, they don't always have all the answers.

Woodlots, aside from their incalculable worth in helping maintain climate stability, protecting stream flow and providing what Purdue University researchers call our most energy-efficient building material, offer farmers profitable markets for veneer, furniture- and construction-grade lumber, firewood, pulpwood, locust fence posts and oak stave bolts.

Knowing all the markets for trees lets a farmer get top dollar for logs. Not taking advantage of those outlets makes no more sense than the greenhorn splitting up good black walnut for stovewood.

Veneer

To qualify for the Cadillac of timber markets, a log must be of exceptionally high quality, free of knots or limbs, straight and straight-grained, at least 8 feet long, without damage from wind, lightning, disease or rot. Veneer buyers won't even look at a fencerow or yard tree for fear the trunk contains metal that could ruin a veneer saw blade, worth many times the value of the log. Black walnut, pecan, white oak and ponderosa pine are often in high demand. There is some call for red oak, oak, hickory and hackberry.

Even in the veneer market, the astute woodlot owner can occasionally find hidden dollars. Some rare woods become stylish and create their own demand. A prime persimmon log, for example, can be very valuable for veneer. Not every timber buyer bidding on your marketable wood is going to be honest enough to admit that if you don't already know it.

Another kind of "hidden" profit in timber applies to all logs, but particularly to veneer. Some buyers and sawyers prefer smaller trees, because the logs are easier to handle. They will try to talk you out of a walnut of, say, 18 inches in diameter. They may argue that a windstorm could ruin it for veneer at any time, so why take a chance on letting it grow? A forester might agree, pointing out that the tree's removal will release seedlings to grow faster to replace it.

If that tree is still growing vigorously, though, consider the alternatives before you saw. Think about how a tree grows. Every year, it puts on about the same amount of diameter. But the larger the tree, the greater the increase in circumference, though diameter growth is the same. A year's growth on a 12-inch diameter tree is much less than half the year's growth on a 24-inch diameter tree. Money invested in a stand of 18-inch walnuts growing to 24-inch size increased much faster than money invested in 12-inchers growing to 18-inch size.

"That's why I'm not in a hurry to sell a good tree even when I've got a willing buyer," says Milton Harman, a tree farmer in Crawford County, Ohio. "If the tree is on a site where it will continue to grow vigorously for five or 10 more years, I'll gamble on waiting."

Veneer buyers do their own cutting. An amateur might fell the tree improperly and ruin it for veneer. No lumber, but especially not veneer, should be cut in the growing season when sap is circulating through the tree. Also, a veneer walnut log needs to be taken out of the weather immediately. Rain can wash the dark pigment and stain the wood, ruining it for veneer.

Furniture-Grade Wood

When it comes to the furniture market, a good, saleable log needs to be almost as perfect as a veneer log. Black walnut, wild cherry, pecan, sugar maple, and sometimes butternut usually command relatively higher prices than yellow poplar, birch and sweetgum. Pine furniture has been in and out of vogue for decades. Now, white oak and even red oak have become very popular.

HIGH-VALUE CROPS AND ENTERPRISES

There is more value in woodland than first meets the eye.

Farmers with time to spend in their woodlot in winter can often discover hidden dollars in furniture wood by selling directly to hobby woodworkers and professional cabinetmakers. For example, a windstorm blew down a smaller, partially decayed walnut in my woods. Timber buyers would not give me the price of firewood for it, yet I was able to cut out more than 100 board feet of good, solid wood in 4-foot and 3-foot sections, worth more than $1 a board foot to woodworkers.

"Yes, quite a few woodlot owners are doing this," says Mark Ogg, a timber buyer and sawmill owner at Wharton, Ohio. "Some farmers get serious about it and invest in a small sawmill. More bring their logs to a commercial sawmill like ours. We do this kind of custom work at our mill by appointment in the evenings and on Saturdays."

Byron Kent, of the Ohio Division of Forestry, has some advice on milling your own lumber for woodworkers. "The smaller, movable sawmills work fine for this kind of work on logs of 12 inches on up to maybe 20 inches in diameter. They don't handle larger logs over 24 inches in diameter very well," he says. "If you decide to mill your own lumber with one of those chainsaw outfits, you need a big, industrialized model, a 50 cc engine, anyway. You should use a special chain with square teeth, rather than angled teeth made for ripping with the grain."

Marketing fine furniture woods yourself is another way to collect hidden dollars. Occasionally, a maple log turns up with either bird's-eye grain or what is variously called tiger stripe, fiddleback, or curly maple grain. Both increase the value of the log considerably. "You can spot the characteristic bird's-eye grain if you peel back the bark and study the surface of the wood closely," says Kent. Sometimes the design is not clear, but you will see the tell-tale sign of it, a pockmarked surface. Curly maple is not visible until you cut into the wood. If you are splitting a piece of maple and it cleaves apart not into two smooth planes, but into rough, washboard faces, you're looking at curly maple. Sand down the washboard surface and one of nature's most beautiful sights in wood grain appears. It is much sought after for gunstocks and musical instruments.

Walnut burl, the lumps of woody growth on walnut trunks, is also highly prized by woodworkers. Fruit woods, especially pear and apple, but even peach, can be sold profitably through woodcrafter magazines.

Construction-Grade Lumber

The softwoods make up most of the commercial trade in this category, served by the large timber growers of pine, spruce and fir. But the farmer with a small woodlot can save himself a significant amount of cash by using his construction-grade wood to build his own buildings. Hardwoods are not easy to work with, but are adequate in barn construction. They can be used before being completely dried out and are still able to take a nail without it bending. A little shrinkage is usually tolerable in barns. The hidden value comes from the difference between using homegrown lumber and buying from the lumber yard at retail prices. Don Reinhart of Alvada, Ohio, says he saved $70,000 by using his own lumber to build his hog buildings. Also, in do-it-yourself construction you can use woods that have little or no market demand. Beech, which few sawmills want to buy, makes good flooring, joists, and plates, nonetheless.

Pulpwood

George Kessler and Jack Cody, Extension forestry specialists at Clemson University, believe woodland can be a better investment than bonds. Basing their calculations on a 5-percent annual increase in the price of pine pulpwood and a median growth rate in the South of 24 cords per acre in 15 years, they say that $1,000 invested in pine trees would give a 15-year, after-tax return of $7,269, nearly three times that of the same money invested in 8-percent bonds.

To increase pulpwood production, foresters are experimenting with an old-world idea, coppicing. With this technique, some trees after being cut send up several new shoots, which grow faster than the original stem, and are periodically harvested. "We've found that sycamore grows very well under coppicing management," says Ted Walker, a forester at the University of Georgia. "Trees are planted close together. The first harvest comes four or five years after planting, and then a stand is cut every other year. We can get up to 8 tons of dry wood per acre per year, and improvements in genetics and cultural practices could conceivably boost yields to twice that."

Specialty Wood Markets

Where good white oak grows, farmers can often sell it for stave bolts used to make whiskey barrels and get a higher return than the tree would otherwise yield. "We look for high-quality logs," says Jerry Bollinger of Vinco Stave and Head Co., Upper Sandusky, Ohio. "Often, we buy the top log above a butt veneer log on a tree. Any species of white oak works for us — bur oak, swamp oak, whatever. Normally we can pay more for stave bolts than sawmills pay for furniture oak."

Fence Posts

Harvesting a farm woodlot for veneer, furniture and other specialty wood markets provides a good second income, but the hidden worth of a woodlot doesn't stop at the sawmill. Young, straight trees and wood fit only for the fireplace are really cash crops that come complete with regular markets and eager customers.

Growing fence posts, for example, is especially lucrative, because the woodlot owner can do all the work himself without expensive machinery, and sell his product at retail. Since good fence posts at farm supply stores cost close to $4 and cornerposts cost up to about $12, you can make good winter wages splitting out posts to sell at even half that price. Black locust, osage orange, red mulberry, red cedar heartwood and catalpa are about the only woods that will last long enough without being treated to make worthwhile posts.

"We think it really pays to grow black locust," says Michigan farmer John Yaeger. "It's easy to start from seed, grows fast, makes a high-BTU firewood, a long-lasting fence post, good bee pasturage when in bloom, and is an impressive soil-improver since it, like other legumes, fixes nitrogen in the soil." Black locust is not advisable in pastures, though. It contains a toxin, robitin, that can cause sickness and occasionally death in livestock.

Yaeger maintains two stands of locust, one spaced in rows 8 by 8 feet or 6 by 12 feet for fence posts, and another for firewood in rows 8 by 12 feet or 10 by 10 feet. It takes locust about 15 years to reach fence post size, so the stand is kept in a 12- to 18-year rotation. He harvests the firewood stand by coppicing. "Coppicing gives several times the yield of firewood as conventional woodlot management," Yaeger says.

Selling pines for utility poles is a specialty that may or may not pay. "The 60-, 70- and 80-foot trees bring a real worthwhile premium for poles," says Ted Walker, a University of Georgia forester. "But there aren't many trees that size, nor is the demand so great for them. If you start selling your 35- to 45-foot trees for poles, the price isn't that much better than saw timber pieces. And if you sell off these trees for poles, you lower the value of your stand to the saw timber buyer to the point where you might not come out ahead."

Firewood

The first hidden value in firewood is in the act of cutting it. If you harvest firewood with an eye to good timber stand improvement (TSI), you are vastly improving the value of your eventual product. In most businesses, that kind of labor and time would be pure overhead, but for the tree farmer, it becomes a modestly profitable sideline.

You can get all kinds of TSI information from your forester, but when you go to the woods with your chain saw to harvest firewood, that information doesn't look as easy to follow as it does on paper. Byron Kent of the Ohio Division of Forestry also has a firewood business. He explains simplified pruning, culling and thinning this way: "The soil of your woodlot can grow only so much wood. Try to get it all on good trees." Crooked, sick and other trees without commercial value become firewood.

HIGH-VALUE CROPS AND ENTERPRISES

Many woodlot owners aren't totally ruthless when improving timber stands, though. Instead, they leave some hollow or other "worthless" trees for wildlife like raccoons, opossums, squirrels and bees. They won't cut down all dead snags, preferring the presence of beneficial woodpeckers drawn to them. They leave some wild fruit trees for deer and birds (a deer can replace $300 worth of purchased meat on the table). Dogwoods and redbuds are left just for pretty. A few dying elms are tolerated for the chance of finding morel mushrooms under them. A choice hickory is often given extra room so it will bush out and produce lots of nuts.

When a stand of young, healthy trees 4 to 8 inches in diameter gets too dense, thin to one tree about every 15 feet. More firewood. And when a tree reaches its prime and is cut for timber, the crown limbs make still more firewood. The harvest of crown limbs is another value not always reckoned by the sharp money boys. Yet, according to a forestry rule of thumb, a 16-inch diameter tree has about one-fourth of a cord of limb wood; an 18-inch tree about one-third of a cord; a 24-inch tree a half cord; and a 28-inch tree three-fourths of a cord.

Kent sells firewood for $105 a cord in the Columbus, Ohio, area. "I don't like to haul over 20 miles. We allow 10 minutes to unload. If it takes much more than that because we have to carry the wood some distance from the truck, we charge $15 an hour for the extra time involved. If customers understand that ahead of time, they don't object." Although he hasn't yet mechanized his loading and handling of firewood, Kent points out that large volume selling—"the only way you can make an appreciable amount of money in this business"—increases efficiency considerably. "Some firewood sellers use a corn hiker to load dump trucks with four to five cords to speed loading and unloading," he says. Others, like Russ Garber, build wooden pallets that hold a rick of wood each and can be moved with tractor hydraulic lift onto and off pickup trucks.

"We've found the rick, which is one-third of a cord, to be the best way to sell wood," says Kent. "That's a stack of 16- to 18-inch-long pieces of wood 4 feet high and 8 feet long. That's the way most people stack wood." He likes to offer a variety of thicknesses in stovewood. "The average piece is about 3 inches in diameter, but we include some up to 6-by-6-inch chunks. Customers like them for night logs."

Another source of hidden profits is double cropping. Natural stands of pecan are managed for nut crops in Kansas, Missouri and other suitable climates. When the trees mature, they can be sold for timber. Maples managed for syrup production eventually become firewood and lumber.

"We suggest to farmers in the pulpwood business that they can make extra money the five years previous to harvest by collecting and selling turpentine gum from slash pines to the naval stores market," says Walker in Georgia.

Dr. Paul Roth, professor of forestry at Southern Illinois University, has completed a 10-year study of multicropping trees with hay and other cultivated crops. "Much of our woodland is a combination of trees and weeds, so why not trees and some usable crop?" says Roth. He found that walnut grown for nuts and/or veneer could be teamed profitably with hay crops on land not entirely suitable for grain. "We have what we call small bottoms in southern Illinois and Indiana—very fertile ground but hard to move big machinery in and out of," he says. "These bottoms often flood in the spring enough to hurt grain products, but not hay crops. Walnut trees and certain grasses and legumes work out well in these situations." Timothy, *Lespedeza sericea*, bluegrass and sweet clover performed best. "If you use the wide tree spacings for nut production—40 to 50 feet between rows—you get the best response from the cultivated hay crop between them," he says. "Then if you prune the trees for a clear veneer butt log, you add the third and ultimately most profitable crop. Pruning limbs off 10 to 12 feet above the ground makes it easier to mow and move other machinery around the trees, too."

Some grasses, particularly fescue, inhibited tree growth in Roth's test. "You must match tree to plant, and both to soil type," Roth cautions. "You don't want to plant walnut on high, thin land. White oak, northern red oak and tulip poplar do much better there."

Just as there are hidden values in woodland to make it compete with cash grain alternatives, there are hidden costs in converting woodland to cultivated fields that are not always taken into account by the farmer eyeing a foreign grain market. "Sometimes I can talk a farmer out of bulldozing his woodland just by comparing costs on less than prime corn land," says timber buyer and sawmill operator Mark Ogg. "It doesn't cost you $100 an acre every year in fuel, machinery, labor and chemicals to grow timber like it does grain. In fact, timber is practically free in terms of out-of-pocket costs. Also, it's becoming very costly to clear land. To do it right and tile it, you're talking about over $1,500 an acre. Even at 10 percent interest, that's $150 per acre per year of hidden cost. I can sympathize with the farmer clearing land in a frantic effort to keep up his cash flow, but on a lot of this second-rate land being cleared now, I know trees will make more money."

Money-making stands of timber don't just happen on their own, though. In upgrading a poor woodlot without good seed trees in it, or in converting fields to woodland, natural reseeding can't be depended on to give you the best stand of trees. Start shallow-rooted trees from seed or look for inexpensive transplants at state nurseries. Often you can find good, wild seed or seedlings available for the asking in nearby woodland. Maple and ash are quite tolerant of shade and grow in great numbers where there are seed trees to propagate them. Maple is particularly easy to transplant. In early spring, you can actually pull up small maples without digging and successfully move them to another site. Small wild cherry trees are easy to move too, but birds usually seed more than you want by eating the fruit.

Deep-rooted oaks and nut trees are best started from seed. In late fall or early winter, simply drop the seeds on the soft ground and gently push them into the soil with your heel. Where squirrels or other rodents are a problem, keep the seed in some cold place over winter and plant in late spring. Oaks—especially white oak, one of the more valuable hardwoods—do not grow well in forest shade. Planted in a forest opening with partial sunlight, they are apt to be crowded by more shade-tolerant species like maple. Try to get oak growing in sunny places before the other species get started. If you are reforesting open land, plant oaks several years before other trees.

White oak grows well where black walnut will not. Plant the walnut on richer, deeper soil; put white oak on clay hills. Forester Howard Kriebel at the Ohio Agricultural Research and Development Center at Wooster recommends using black walnut seeds that come from an area about 200 miles south of your location, if possible. They will grow faster, according to tests done in Ohio and Minnesota. At the more northerly limits of black walnut hardiness, trees from 200 miles south occasionally might not be winter-hardy enough. So, advises Kriebel, plant a mixture of native and southern seed. In planting black walnuts, you need not remove the husk from around the nut.

John Yaeger, the Michigan farmer who grows black locust for fence posts and firewood, starts his trees from seed. "Our method is simple and straightforward," he says. "The seed can be planted right away, it has no need of after-ripening dormancy to germinate as many tree seeds do. But it does have a hard coat. One procedure is to soften the seed coat by placing the seeds in hot water of about 200 degrees F., then allowing the water to cool to room temperature, the seeds soaking for about 12 to 18 hours. Those seeds that are plump and swollen after this treatment will germinate readily. Another method is to nick the seed coat with a sharp knife, then soak the seed in water at room temperature overnight.

"You can seed directly into a permanent location, but losses will be high. We start seeds in flats or pots about half an inch apart and about one-quarter inch deep. Germination occurs in about six days when the soil is 70 degrees F. We transplant the seedlings to 2-inch pots in about two to three weeks, and then set them out after another three weeks, hardening them off before they go to the nursery row or permanent location. If in a permanent location where we can't cultivate as throughly as in the nursery, we use a heavy mulch or hoe around the little trees occasionally."

Small amounts of locust seed are available from Mellinger's, 2310 W. South Range Road, North Lima, Ohio 44452. Mellinger's sells a variety of tree seeds. So does Redwood City Seed Co., P.O. Box 361, Redwood City, Calif. 94064.

Another way to sweeten your woodlot income is by planting so-called "super maples" developed by Ohio foresters. Their sap contains nearly twice the amount of sugar as ordinary sugar maples. Seedlings are being sold as they become available from state nurseries. Write: Ohio Department of Natural Resources, Division of Forestry, Fountain Square, Building C, Columbus, Ohio 43224.

PYO Wood Earns $9,000

DAVE McCOY

FREDERICKTOWN, Ohio—When I was trying to break into regenerative agriculture in 1980, I stepped back to take stock of all my farm's resources. I wanted to implement intensive, rotational grazing, but much of my farm's original pasture was very wet and overgrown with brambles, cattails, and assorted hardwood trees. What a mess. A bulldozing contractor offered to clear the largest area for $8,000 to $10,000, which I didn't have. That's when I came up with the idea of pick-your-own firewood as a money-maker. It's one idea that could be used by other farmers needing cash to invest in improving their land, too.

I placed an ad in the local newspaper, offering pick-your-own firewood at $10 for a small pickup load, $15 for a regular pickup (half or three-quarter ton), and $20 for bigger loads. Around here a cord of split wood runs $60 to $100.

After three years, we're approaching $9,000 in firewood sales that have paid for two-thirds of the bulldozing needed to level and ditch the wet areas, all cleared by local wood burners. My goal is to have the project pay for itself even as we reap benefits from better use of the land.

This firewood business comes with its share of hassles, but we feel it's worth it. For a reasonable price, we've helped heat hundreds of homes owned by people willing to do the work themselves. We had to increase our liability insurance, spend money on advertising, and buy better chain saws to cut trees in areas too wet for people to get to. We've had to keep a tractor handy to pull out people's vehicles that got bogged in the soft spots, and accept the fact that some people will leave without paying. I learned that hunters and woodcutters just don't mix, and that there are some real characters out there, like the ones who won't cut where you direct them, and the guy who pulls up in a Ford Escort with a handsaw.

This all calls for working Saturdays, burning branches, and putting in extra hours in bad weather, because I can't let my herd and crops suffer while I'm away cutting wood. But the time and trouble has yielded us another 15 acres of expanded, renovated pasture. When we're finished, we'll gain a few more acres for field crops. We hope good pasture management, combined with a little chain sawing of stray saplings now and then, will keep back encroaching trees at the woods' edge. Keep in mind that we have yet to cut a single tree in our 45 acres of established woods; we hope the forester will show us how to manage these trees so the good timber will serve the next generation.

Dave McCoy milks 45 cows on his family's 205-acre farm near Fredericktown, Ohio.

Chapter 6
Add Value To Everything

20 Times More Income From Wool
With a cottage industry-scale wool processing machine that paid for itself in just seven months.

WOMELSDORF, Pa.—Like all too many wool producers, Donna Kennedy used to sell raw wool to the wholesale market for a mere 47 cents a pound. But now, she's tapping markets that pay $10 per pound more. The key is a new—affordable—cottage industry-scale, wool processing machine that makes Kennedy's wool a real value-added product, increasing its worth at least twentyfold. In addition, Kennedy earns extra income with the machine by doing custom processing for other shepherds.

"It's really turning the farmstead around, making it an income producer for us," says Kennedy, who, in partnership with her husband and another couple, farms 15 acres and runs the wool processing business.

By taking raw, often dirty and tangled fleeces and working them into clean, usable form, they're meeting the needs of the growing handspinning and crafts market. Although this may seem like a specialized niche, the market has plenty of room for growth. Domestic producers aren't even close to meeting the demand. The United States imports more than 70 million pounds of wool annually. Tens of thousands of pounds goes to the handspinning and crafts market, according to USDA's Foreign Agricultural Services.

Once a wool processor establishes a reputation, work usually comes pouring in. "We received the machine in March and we were going full tilt by April," says Kennedy. "We processed 2,000 pounds of wool by early December." Kennedy handles all the wool from the flock of nine sheep on her Amazing Acres farm. The bulk of her work comes from neighbors and from people who see the farm's ads in sheep trade magazines. While charges for custom work vary, Kennedy says such orders bring in $700 per month. At that rate, the $4,900 investment (including freight) in the machine was recovered in seven months.

Become Self-Reliant

For Kennedy, the idea of processing her own wool and doing custom work developed as a natural extension of her interest in sheep, wool spinning, weaving and knitting. She knew that good, clean carded wool was valuable and sometimes hard to get, so why not produce her own and extra to sell? Her interests led her to visit a large wool-processing plant in Vermont. "I learned a lot, but it was discouraging at the same time," she says. Their wool carding machine was big, complicated and cost $30,000. "You practically have to be a master mechanic just to run it," she says.

Taking fleece from its raw state, right off the sheep, to a condition that permits the wool to be spun into yarn or used for other crafts is a four-step process. First, the wool is skirted — trimmed around the edges to remove small pieces of soil, hay, burrs or dung imbedded in the fleece. The skirted fleece is then washed in water and detergent to remove any remaining dirt and most of the natural oil, called lanolin. The wool is then teased or fluffed with a picker, a simple, toothed device.

So far, so good, for the do-it-yourselfer. But now comes the bottleneck: carding. In its simplest form, carding is the repeated brushing of the wool to bring it to an even consistency and shape. Hand carding takes so much time, it's unprofitable for processing any real quantity of wool.

Enter Patrick Green Carders Ltd., of Sardis, British Columbia, Canada. "Before our cottage-scale carder, there was no machine to fill the gap between hand carding or the table top drum carder and the large, expensive mill machinery," says Patrick Green, who came to Canada from England in 1963.

His business was born when his wife, an avid spinner and weaver, began looking for faster and better alternatives to hand carding wool. When she found none, Patrick built her a machine and realized that he had also solved a problem for many wool craftspeople. He quit his job "piloting a desk for a big industrial firm" in England, moved to British Columbia, and began selling and developing tools for cottage industry-scale wool work.

Set Your Own Price

The latest model of the cottage-scale carder has been on the market since 1979, and was developed with the help of Paula Simmons, professional handspinner and author of several books, including "Spinning and Weaving With Wool," published by Garden Way. Simmons says the machine is useful for those who specialize in making quilt batts and quilts; co-ops doing cooperative carding and marketing; specialists in hand-dyed, handspun yarns that can blend any colors they choose in the carder feed tray; Angora goat raisers, who can increase the value of their already pricy product; feltmakers using wool batts for handmade felt; and sheep-raising handspinners, who card wool for their own use and for sale.

"If you're producing good, clean fleece, don't settle for the wool pool price. Clean it, card it, knit it into a sweater if you have to. It'll be worth your time," says Simmons. "If you don't think your wool is valuable, go to a knitting shop and buy some. You'll pay for it *by the ounce*."

Although the cottage industry-scale carder is just 48 inches high, 60 inches long and 28 inches wide, it weighs 400 pounds. "It's built to last for many years of hard use, and if anything does wear out, it's replaceable with off-the-shelf parts," says Green. The machine uses standard bolts, belts, pulleys, and bearings available anywhere in the United States and Canada.

The machine can turn out roving, a 1- to 3-inch-wide band of even-textured wool used for handspinning, at 1,200 feet (6 to 8 pounds) per hour. It can churn out batts 72 inches long, 20 inches wide and up to 4 inches thick at 8 to 10 pounds per hour. It can handle the long-staple wools used by handspinners from 2 inches and longer, including mohair. The machine has seven carding drums (the "teeth" of the unit), which are 16 inches wide and are clothed with commercial-grade carding cloth.

"My goal is to build it simply, and build it well," says Green. He has sold nearly 30 of the machines, mostly in the United States and Canada, although orders have come in from other countries and as far away as New Zealand.

The carder is simple to operate and adjust, too. "We didn't do any carding before we got the machine, so we had to learn as we went," says Kennedy. "If you really put your mind to it, you could master the basics in a couple of days. It takes about three hours per fleece, from the start to finish, and it's well worth it."

Editor's Note: *Besides the cottage industry-scale carder, Green makes and sells a table top carder, pickers, and hard-to-find wool tools such as doffer sticks, felting brushes and picker claws. For a free catalog, write: Patrick Green Carders Ltd., 48793 Chilliwack Lake Road, Sardis, British Columbia, Canada V2R 2P1. Phone: (604) 858-6020.*

They Spell Profit C-I-D-E-R
Adding a cider press will often triple cider sales.
AUGUST SCHUMACHER

Cider making is a growing, profitable business. People are attracted to cider pressing. They like to see the making of a freshly pressed, nicely blended, properly chilled cider. With an investment of $6,000 for used equipment, a farmer with a few hundred apple trees can press 5,000 to 6,000 gallons from windfalls and second grades. At an average of $2 per gallon, you can convert $1.50- to $2-a-bushel apples into cider worth the equivalent of $7 per bushel. Thus, some 1,850 bushels of apples worth only $2 each at the processor (for a total of $3,700), can be retailed in the form of cider for more than $12,000.

"Most new cider makers should be able to pay for their press within two years," says Norman French, a cider mill manufacturer in Conway, Mass. It is not uncommon for orchards that are selling their windfalls and buying back wholesale cider for their roadside stand to "double and sometimes quadruple cider sales once they install their own press," he says.

Cider maker Arthur Amidon of Wilton, N.H., says he sees a "real boom in this business, with ample room for the small, serious maker of quality, blended fresh cider." Ed Waseem of Milan, Mich., reports he makes more than 40,000 bottles a year. "This cider business has become really remarkable," he says. In the 1960s, there were only 10 or so known cider mills in southeast Michigan. By 1979, there were 118 mills in the same area. "This thing is really booming," Waseem says.

Orchardist Jim Roberts of Cumberland, R.I., presses cider from mid-September to June, using cold storage, and later, apples graded from controlled atmosphere facilities. He also freezes several thousand gallons for customers who like fresh cider so much they insist on year-round availability.

The first step in pressing cider is to build a layer of pomace called a "cheese."

Streamlined jugging methods allow one person to fill 300 to 500 gallons a day.

In California, John and Sandi Dach operate a small cider mill north of San Francisco. Because they use almost no pesticides on their orchard, their cider brings a premium price of $4.50 a gallon. The Dachs also freeze cider for their expanding off-season trade.

Even if a farmer has few or no apples, buying cider-grade apples is still financially attractive, provided most cider sales are at retail prices. Gould Hill Orchards in New Hampshire, like many older cider makers, is replacing its smaller presses with larger-volume presses. This kind of expansion puts reasonably priced used presses on the market for growers who want to experiment with cider sales.

Cider making is especially suited to smaller farms. The technology has changed little since the early 17th century, when the basic grinding and pressing methods were developed. It involves mashing or grinding fresh apples into a pulp called pomace, then building up successive layers of this pomace into a "cheese" for pressing. Apples are about 80 percent liquid, and the grinding into a pomace enables the pressing to squeeze most of this liquid out as the fresh juice we call sweet cider.

Over the centuries, the biggest changes in cider technology have been: (1) the replacement of straw by cloths in building the "cheese", (2) the introduction of the mechanical grinder for pomace making, and (3) the invention of the hydraulic cider press to replace the man-powered wooden screw press.

There have been refinements in recent years, but the basic principles of pomace making and batch pressing are still the same for most cider mills, today.

While large apple processors use huge German and Italian machines to centrifuge apple juice, modern technology does not seem to be able to duplicate the flavor of freshly ground apples layered into seven or eight cheeses and squeezed under 2,000 pounds of pressure for 15 to 20 minutes.

Another reason cider making is so suitable for small farms is the perishability of the product. Made without preservatives and stored in a cooler, cider rarely lasts more than a week. Most good cider makers usually press daily and try not to store cider for more than a day. After three days, they will pour out unsold cider. With such a short shelf-life, the best sweet cider is just not sold in supermarkets. Families who get a taste for quality, blended apple cider will travel weekly to your mill.

With cold storage and some buying from controlled-atmosphere storage, many cider makers offer cider from early September to mid-July. This enables them to generate cash sales in the dead of winter and early spring months. As a result, there are some 3,000 smaller farmers now making cider in the U.S., according to estimates by cider press manufacturers.

How difficult is cider making? The basic principles are simple. The most successful cider pressers chill their cider apples to about 35 degrees F. Apples are normally washed before pressing, though some millers feel this dilutes the cider by as much as 10 percent. While windfalls are commonly used, best results are obtained from tree-picked, sound apples. Later-ripening apples such as MACOUN, DELICIOUS and BALDWIN seem to give a tastier cider than the earlier varieties like MCINTOSH and CORTLAND. The best tasting ciders are often obtained from blending the rarer original cider apples, such as ROXBURY, RUSSET, SEEK-NO-FURTHER, and RHODE ISLAND GREENING, with the BALDWIN and MCINTOSH.

In the East, Norman French likes a MCINTOSH-CORTLAND blend. In the Midwest, Wesley Day, a press maker in Goshen, Ind., suggests a JONATHAN-MCINTOSH-DELICIOUS blend. A straight GOLDEN DELICIOUS or RED DELICIOUS cider is "really too sweet," reports Michigan's Ed Waseem. "You really get your best varieties of apples after about the middle of October. Then we can have consistent quality right through the winter, because we have a lot of different kinds of apples," Waseem says.

Brushed and washed, the apples are then milled or ground into a pulp in a mechanically driven grater. The pulp is let down onto woven cloths laid over a wooden rack. Once filled to a depth of 3 inches, the cloth is folded over in the shape of an envelope and a second rack is added to the stack. The new rack is then covered with a cloth, filled with pulp and again "enveloped." This process is repeated until some six to 10 racks are filled. Once built up, the entire pile, now called a "cheese," is rolled on its own "truck" over the hydraulic ram where pressure is then applied up to 2,000 pounds per square inch. This squeezing results in 3 to 4 gallons from one bushel of apples, or some 175 to 180 gallons per ton of apples. Sometimes the squeezed pulp is resqueezed to obtain the final juices. In France, this juice of the second pressing was occasionally used for fermentation and distillation into apple brandy, which Ernest Hemingway regularly drank as calvados.

This pulp is sometimes spread onto fields as a "pomace manure," or is fed to cattle and pigs. A few cider makers dry the pomace and use it for making pectin.

What about health regulations? States vary in their requirements. Many states have their dairy farm inspectors check cider mills. The minimum regulations usually require hot and cold water, window screens, concrete floors with drains, and ceilings and walls covered with white, washable Masonite sheathing. Aside from pleasing inspectors, good cider makers try to keep their mills spotless, as many customers like to watch pressing, and a dirty mill is unappealing. Massachusetts requires dating of fresh cider on the jug, which adds to the "fresh" image cider now carries.

What equipment is required? A good, used 22-inch press with conveyor, hopper, washer, bulk tank, jugger, racks and cloths probably can be bought and installed for about $6,000. Renovation of a building — installing hot water, buying a washing machine and ordering an initial supply of plastic jugs — will add another $1,500 to $2,000. A new 22-inch press, with all attachments, costs about $12,000. Most cider makers get 70 to 80 gallons per hour from a 22-inch press, which takes two people to efficiently set up, operate and clean.

A new 17-inch press produces 30 to 40 gallons per hour and costs about $2,300. Add another $2,500 to $3,000 for transport, set up, building renovation, tanks, conveyor, pump, and bottler.

A sanifeed system that automatically grinds pomace and feeds it through a hose to the cheese permits one-man operation. With a good, efficient jugger and perhaps some special spray attachments to rapidly clean the press, one person with a little bottling and cleanup help can turn out 300 to 500 gallons in a day. If your volume is at this level, Norman French feels the larger press with sanifeed is definitely worthwhile. The sanifeed package is available from Conway Equipment in Conway, Mass., for about $3,500. It will serve a 28-inch press.

Cider making is fun, unique, and profitable. It is an excellent small business. Well run, a cider mill can add $3,000 to $6,000 a year to your farm's net income.

August Schumacher has raised a variety of crops on his family's truck farm in Massachusetts. In 1985, he became that state's agriculture commissioner.

Editor's Note: *Cider making is one of the perfect enterprises for the small, diversified farm with a Clientele Membership Club. "It's a real attraction to kids and others to see that cider press in operation," says Tom Brewer, an Extension food marketing specialist at Pennsylvania State University. While no statistics are kept on cider or even apple juice production in the United States, Brewer says there is a general agreement that "consumption is high and rising."*

Cider makes the most profitable use of cull apples. More importantly, it helps fill up your clients' freezers, because the only way to enjoy apple cider year-round is to freeze a bunch of it each fall. Just be sure to tell your customers to drink a small glass of cider from each jug before placing it in the freezer, though, to compensate for the expansion caused by freezing. They'll have a sticky mess inside their freezers if they don't.

For information on suppliers of cider making equipment, see the American Fruit Grower. *Every year, the magazine publishes a list of suppliers of everything from corers and pulpers to presses, graters and grinders. Write:*

American Fruit Grower
37841 Euclid Ave.
Willoughby, Ohio 44094

Double Early Produce Prices

The secret is row covers and new mulches of plastic, aluminum and paper. They all raise yields, extend growing seasons and help earn top dollar.

BEDFORD, N.H. — Produce grower Richard Clark is using a new plastic row cover and mulch system to boost spring soil temperatures and fend off late-season frosts that often punctuate New Hampshire springs. And his extra-early tomatoes are bringing up to twice the price of his mid-season crop.

"People just wait for the season to start. If you have the tomatoes, you'll have as much business as you can handle," Clark says. "Tomatoes sell for about 39 cents per pound or three pounds for a dollar in mid-season. Early tomatoes sell for 20 to 40 cents more per pound. It's like sweet corn — you've got to get the early market or you're missing out."

But row covers aren't the only breakthrough. New mulches that look like huge strips of tinfoil or recycled shopping bags are repelling insects, conserving moisture, saving on labor and increasing high-value crop yields across the country.

For example, Lou Galetto, a farmer near Vineland, N.J., is getting profitable fall squash crops, despite a severe aphid problem. His aluminized paper mulch confuses aphids so much they rarely even land on a squash plant.

For Clark, who grows some 70 acres of vegetables, including 6,000 tomato plants, row covers and mulch literally are changing the way he farms. In addition to boosting tomato yields to an almost-unheard-of 40 pounds per plant, the system controls weeds and insects without chemicals.

"The row covers should be thought of as a growth intensifying tool, rather than something that just helps you plant earlier," explains Otho Wells, a University of New Hampshire vegetable crop specialist. Heat-loving vegetables grow vigorously and give high yields when protected with row covers, he says. Muskmelons, watermelons and other vine crops benefit from being started under row covers as much as tomatoes and peppers. In fact, in some parts of the North, using row covers may be the only way to raise these crops economically.

Row covers increase early spring soil temperatures six to eight degrees F, and provide 3 to 4 degrees of frost protection. "I leave the row covers in place until the end of May," Clark says. He then rolls the covers back to prevent overheating during warmer weather. The covers are secured by the rows, though, because they will go on again in the fall to extend the season. "We picked our very last tomatoes on November 21," Clark says. We had some researchers visit here in mid-October, and they just couldn't believe the production we were still getting."

Clark and University of New Hampshire researchers find row covers firmly anchored in the soil resist wind and weather damage. And it's not necessary to remove the covers to replace transplants. The work can be done through the slits.

Miniature Greenhouses

Insects and diseases haven't been problems for Clark with row covers. He suspects conditions may be too hot for insect pests under the row covers or that the covers just provide conditions the pests find hostile. But the warmth

An inside look at a slitted row cover. Growth-intensifying tools such as row covers boost yields and prices — and help control pests.

can, under some conditions be dangerous to plants. It can hit 120 degrees F under the row covers on a sunny, 95-degree day, but such conditions are unusual for the times of year the covers are used. Also, open tomato blossoms can be burned if they are touching or too near a row cover on a hot day.

Clark's black plastic mulch and clear row covers effectively trap solar energy to provide "miniature greenhouses" for his crops. But several hundred miles to the south, near Vineland, N.J., truck farmer Lou Galetto is using aluminized paper mulch to reflect as much light as possible. He's not doing it to keep the soil cool, though. The reflected light actually prevents most aphids from ever landing on his squash plants.

"I couldn't grow squash in the fall without reflective mulch," says Galetto, who has been using the mulch for several years. "I can get by without it in the spring, but the aphid populations peak later in the season." He says aphids spread disease, which can wipe out a squash crop in a matter of days.

"Aluminum mulches do produce practical reduction of aphid-transmitted mosaic viruses of squash and thrips. Leafminers were also repelled when aluminized mulches were used with gladiolus and squash," concludes a USDA study of reflective mulches in Charleston, S.C. "On the other hand, honeybees and pickleworm were attracted to squash plants mulched with aluminum."

Reflective mulches generally do not discourage pollinators and other beneficial insects, researchers say. And the mulches are effective on cloudy days, as well as sunny days.

Complete Weed Control

About 100 acres of squash and Chinese cabbage in the Vineland area are mulched with aluminized paper at the fall peak, says Jeff Johnson, Extension agent for Cumberland County, N.J. Area truck farmers, many of whom have 10- to 20-acre farms, rely on reflective mulch for more than just insect control, he says. Since no light penetrates the mulch, it gives complete weed control. And the mulch's moisture-retaining quality helps tide thirsty plants like squash over some dry spells.

All this makes aluminized paper more than worth its $93 per roll cost, Johnson says. About four and a half rolls per acre are needed for squash.

Cuts Labor Costs

Also, there's no labor cost involved in picking up used mulch, Galetto says. It is disked into the soil once the season is over.

The labor costs and trouble of picking up some mulching material is a big reason why few farmers choose mulch over herbicides. At season's end, mulch is usually covered with matted, rotting vines and fruit, anchored firmly in the soil, and almost impossible to remove easily.

But a new, biodegradable paper mulch puts an end to such headaches, while providing all of the benefits of an opaque mulch. Gardeners have used paper mulches for years, but there was no practical paper mulching system for farmers. Paper mulches can be heavy, and difficult to apply without tearing. And in some cases, the paper deteriorates before its job is done.

Now, durable paper mulches and mulch-laying machines are available from Orebro Papersbruk, a Swedish firm, which makes Hortopaper brand mulch from cellulose and peat fibers. "Tomato yields increased 80 percent with the use of this paper mulch, with respect to the tomatoes grown without it. When the tomatoes were graded, the marketable yield differences between the paper mulch-grown tomatoes and the untreated ones became even greater. Not only were there total and marketable yield differences, but also size differences were registered," says Pedro Ilic, Extension farm advisor for Fresno County, Calif., where the paper mulch was tested.

The dark brown mulch also raised soil temperatures for a season-extending effect. The mulch is being tested with newly planted fruit trees as well as with vegetable crops. The 24-inch wide paper costs 6 cents per foot; 32-inch paper is 8 cents per foot, and the 48-inch paper is 12 cents per foot. For more information, contact: Actagrow, 4111 N. Motel Dr., Suite 101, Fresno, Calif. 93722. Phone (209) 275-3600.

"Many types of mulch are cost effective," adds vegetable specialist Wells. "Sometimes, just a week of earliness can pay for the mulch, not to mention the yield increase, weed control and moisture retention. I think we'll see more growers using row covers and other mulches."

Clark's Row Cover/Mulch System

1. Lay 4-foot wide black plastic mulch.
2. Plant seeds or transplants 10 days to two weeks earlier than usual.
3. Cut 63-inch hoops of No. 8 or No. 9 wire. Install hoops at 4- to 5-foot intervals along bed by pushing ends of hoops six inches into soil.
4. Dig a 3- to 4-inch deep furrow on each side of row, outside hoops.
5. Stretch clear plastic over hoops. Secure in furrows with soil.

The row covers, which are really miniature solar greenhouses, are now in place and will need no further maintenance. Clear plastic row covers come pre-slitted for proper ventilation. One supplier is Ken-Bar, Inc., 24 Gould Street, Reading, Massachusetts, 01867. A five-foot wide, 500-foot long roll of slitted row cover costs $36. Orders of more than 50 rolls receive a discount.

Cash In On Raw Goat Milk Cheese
Wisconsin co-op finds dairy goat products profitable and popular.

SOLDIER'S GROVE, Wis. — The 547.4 million pounds of cow cheese lying unsold in government warehouses doesn't worry a young Wisconsin dairy co-op one bit. This co-op milks goats. And while demand for cheese made from cow's milk is so low that USDA gives away millions of pounds of its surplus cheese, annual demand for the co-op's raw goat milk cheese has jumped from 600 to 75,000 pounds in recent years.

The 20 members of Southwestern Wisconsin Dairy Goat Products Ltd., selling under their Kickapoo of Wisconsin label, receive milk checks $5 a hundredweight fatter than neighboring cow dairymen. The co-op commands wholesale cheese prices nearly double that of cow cheese.

"There's a real demand for this product out there," says Orvin Buros, a cheesemaker for 49 of his 66 years, who helped the co-op perfect its cheesemaking operation. "And these small farmers milking goats got something over the big boys. They don't need to borrow that big interest money to buy all that fancy equipment and silos. I would go with goats today if I wanted to milk."

Co-op members don't fit the stereotype of goat dairies, though. Only two of the 20 members still milk by hand. Some have bulk tanks and pipeline milking systems that are standard on any commercial dairy. Most herds have from 20 to 80 milkers, but co-op members Jim and Shelly Johnson of Ferryville milk about 150 goats with the help of their four children. Beuford and Sharon Foley, who have a 40-acre farm near Highland, milk more than 250 goats in a double herringbone parlor with eight units. Each milking takes about three hours.

The Foleys, both of whom grew up on dairy farms, say they wouldn't want to go back to milking cows. "Goats have a personality all of their own. They're a lot safer to be around, too," Foley adds. "You don't get a broken arm when a goat kicks you." The Foleys buy about 95 percent of their feed. Their 40 acres are devoted to pasture.

The Johnsons, on the other hand, raise all the baled hay and ear corn their herd requires on their 200-acre farm. They are the only co-op members dairying full time. The family also has a 10-acre allotment for raising chewing tobacco.

Johnson, a former civil engineer, and his wife, who started raising goats as a backyard hobby, blocked off a front section of their old dairy barn to create a milking parlor for

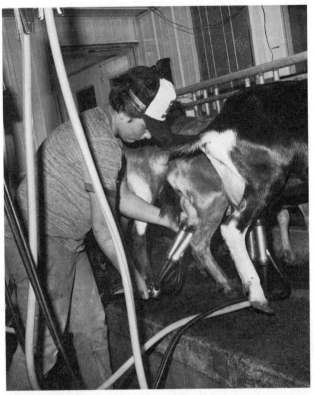

It's not only easier to milk goats than dairy cows, it can be a lot more profitable.

their goats and gutted the rest of the building to form a large free-stall area. A new pole barn houses young stock.

Production Rising

The Johnsons' goats average 1,350 pounds, or about 5 pounds of milk per animal every day of the 305-day goat lactation period. The Johnsons believe a 2,000-pound average is a reasonable goal for their animals. They crossbreed high-producing Saanens with Nubians, which give a richer milk. The herd's butterfat average is 3.5 percent. The couple feeds a mixture of coarsely ground shell corn and oats, plus a commercial dairy supplement to boost protein content to 14 percent. Lactating does receive about 4 pounds of grain a day.

The common complaint of co-op members is the lack of upbreeding to increase the milking potential of goats. Because of this country's infatuation with the dairy cow, little research has gone into goats, they say. The dairy goat today is where the cow was 30 years ago. Artificial insemination for goats is just now coming into practice.

Despite that, the goat herd owned by Jim Luti and Carol Bona of Arena already averages more than 2,000 pounds per animal. Most co-op members are replacing poor producers with younger, carefully chosen animals that are expected to increase production averages. A common grade doe owned by the Foleys, for example, produces 3,500 pounds of milk.

$5-Per-Cwt Bonus

Good dairy goats now sell for about $100 each, according to Mike Hankin, who founded the co-op and is now its licensed cheesemaker. On paper, he says, a herd of 100 milking does giving 2,000 pounds each could gross $42,000 a year. The co-op sets its own price for milk and pays $5 more per cwt than cow dairymen receive, Hankin adds.

Despite that, Hankin doubts that anyone milking fewer than 50 animals can make money with dairy goats. About a dozen area producers who were buying feed for smaller herds have dropped out of the co-op. "When they figured out the money and time they were investing, it just wasn't a worthwhile investment," he says.

Hankin was milking 45 goats when he persuaded a cheese factory in nearby North Clayton to make a vat of goat milk cheese every five days in 1974. That first year, 600 pounds of cheese were produced. In 1975, he lined up five other goat milk suppliers and formed the co-op. Hankin became an apprentice cheesemaker at the North Clayton plant in 1977, and reluctantly sold his herd to concentrate on cheesemaking. A year later, he had his cheesemaker's license—a legal necessity in Wisconsin—and has done all the co-op's cheesemaking and marketing ever since.

"The small cheese factories, the older cheesemakers who helped us get going, the amount of used, small-scale dairying equipment around at reasonable prices, these and other factors were all in our favor," he says.

Goats are also well-suited to the rough land in southwestern Wisconsin. Like deer, they actually prefer brush to lush pasture. Goats do well where cows won't. Co-op members say their goats clean up practically everything but walnut trees, leaving only mature trees behind in their woodlots. As a result, serious overgrazing is something they must constantly guard against.

"Goats are really quite timid animals who follow a leader in their herd. They don't like to wander too far from home," Hankin adds. Most co-op members confine their animals with 3-strand electric fence. A 32-inch woven-wire fence topped with a single strand of hot wire works even better, they say.

The co-op owns its own bulk truck, which hauls milk within a 100-mile radius of the cheese factory. Cheese is made once every three days, from March through November. In addition to state tests of bacteria and sediment levels, the raw milk is tested for pesticide residues, guaranteeing customers like Shiloh Farms a pure product.

Hankins says the 40-pound blocks of raw goat milk cheese sell for $2.40 a pound, compared with $1.30 a pound for regular block cheddar. The goat milk cheddar retails for from $3.50 to $5 a pound. Hankin believes the co-op's main customers are gourmet cheeselovers, the health-minded and those allergic to cow milk products. Never heating the goat milk past 96 degrees F preserves all the vitamins, minerals and lecithin, he adds. Goat milk cheese also contains 20 percent less salt and half the cholesterol of cow milk products. No colorings or preservatives are used.

International Market

About 40 percent of the co-op's cheese is marketed through IntraCommunity Cooperative (ICC), which is located about 70 miles away in Madison. ICC supplies food co-ops throughout the nation. Another 40 percent—about 2,500 pounds of cheese per month—is sold to Alta Dena, the giant raw-milk producer in California. The cheese is marketed under the Health Valley label. The rest of the co-op's cheese is sent via UPS to stores and individuals on special order.

In addition to cheese, the co-op ships a semi-load of kid goats to New York City every year. The kids are bought mostly by Greeks for their traditional Easter meal. Goat meat, called chevon, is especially lean and tasty, and quite popular with Mexicans, Filipinos and other ethnic groups.

'Little House In The Woods' Is Retreat For Vacationers

This 19th century, Victorian bungalow can earn up to $1,000 a month as a vacation retreat.

ELKADER, Iowa — Some vacationers like solitude, while others prefer bustling, highly commercialized playgrounds where they are waited on hand and foot. For those vacationers who want to step off the beaten path and experience relaxation in the country, "Little House in the Woods" has the right mix of activities at an affordable price.

Little House is the brainchild of Pat and Milford "Muff" Koehn. This farm couple turned a run-down, 19th century farmhouse into a vacation haven that is booked 30 weeks a year. Little House sits on land that the Koehns added to their 1,000-acre farm in 1974. The Koehns rent the house for $100 a weekend or $250 for one week. Little House is set up to accommodate eight people, but Pat says there have been as many as 12 adults staying in the house.

"When we bought the farmland, I fell in love with the house," Pat says. "Rather than tear the house down, we decided to fix it up and try to rent it to vacationers. The initial idea was to have something similar to a dude ranch. But the house has developed into something a little different than a dude ranch, because most of our vacationers prefer privacy. Some families bring their children to our farm where they can feed calves, but that activity grows old pretty fast.

"Even though we didn't know other farm families who were renting to vacationers, we were confident Little House would work. We thought our area had a lot to offer vacationers, and we thought people from urban areas would appreciate our relaxed rural setting. We are not only selling a place for people to stay for the night, but we are also sharing our rural lifestyle. People obviously appreciate what we offer, because we have many repeat customers. We've also made a lot of new friends, which makes the project more fun for our family."

Pat says the chances of renting Little House on a monthly basis were not good when they bought the house. People weren't interested in living in the country when there was rental property available in villages and towns. So the Koehns decided that more income could be earned by renting the house over weekends or by the week.

Pat Koehn says Little House in the Woods offers vacationers privacy in a relaxed rural setting.

Retaining Country Charm

One of the Koehns' goals when they started refurbishing Little House was to retain as much of its 19th century charm as possible. The house was constructed in 1891 by a Norwegian immigrant. The house is a 3-story, Victorian bungalow. The Koehns spent about $8,000 to get the house in peak condition. They sanded and finished all the hardwood floors, installed new plumbing, constructed a fireplace, replaced the roof and completed many minor repairs before it was opened for vacationers.

During the restoration project, the Koehns relocated the kitchen from the lower level to the second story. The first level is now used as a family or living room and has a large fireplace. The room also has a couch that folds into a bed for two people.

"The wooden floor in the old kitchen rotted away, so we replaced it with a concrete floor and added the cozy fireplace," Pat says. "The fireplace may not be authentic, but we felt the cozy flicker of a fire and the extra heat were reasons enough to do the work."

The furnishings inside Little House are very basic. Pat is constantly upgrading the furniture, linens and dishes, especially when she sees something in local shops that was made during the late 19th century and is reasonably priced. In the future, Pat would like to market Little House as an authentic 19th century farmhouse in an effort to increase bookings. Since Pat and Muff opened Little House, many local parents have reserved it for weekends to show their children a kind of homestead that was built in the late 1800s.

Generating Publicity

Deciding to fix up the house for vacationers was an easy step, according to Pat. The real challenge was getting the word out so people could learn about the Koehns' Little House vacation spot.

Pat tried advertising in regional, daily newspapers, but the response was minimal, and the ads were too expensive to continue on a regular basis. The ads also could not highlight all the activities available near the Koehns' farm. Many people who inquired about the house expected it to be on a lake or a river. The Turkey River runs less than a quarter of a mile from the house, but people were more interested in being closer to water. Renting got off to a slow start.

Then, about one year after the house was opened to vacationers, *The Des Moines Register* ran a series of articles on inexpensive vacation spots. Pat saw this as an opportunity to get free publicity, and it worked.

She sent a folder of information to the newspaper, and one columnist featured Little House. Pat says by noon the same day the paper was printed, she had received five calls asking for more information.

"If we were starting over, or if someone would ask me for advice in starting something similar, I would notify vacation or travel editors of large newspapers," she says. "A few well-placed stories in regional newspapers would have gotten us off to a faster start. We should have also put all our advertising money into a large city newspaper, instead of placing ads in several smaller newspapers.

"I'd also place an ad large enough to mention all the activities available to people visiting our area. I'd highlight fishing on the Turkey River, hiking in our surrounding woods and shopping in nearby village stores. We'd also direct the ad to people who travel far for vacations."

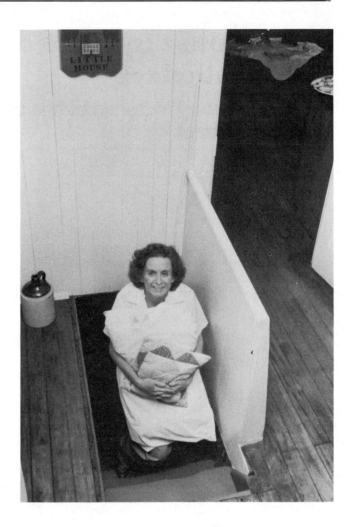

Selling Local Attractions

Pat and Muff knew their Clayton County home would attract vacationers. This helped them justify going ahead with the Little House project. They say some rural residents overlook or take for granted natural attractions and other interesting local sites or shops that vacationers appreciate.

Clayton County, for example, is part of what local residents call Iowa's "Little Switzerland." There are thousands of acres of lush hills to hike on during summer or ski on during winter. The Turkey River that runs through the Koehns' property is stocked with trout, and it's ideal for swimming, canoeing or inner-tubing. Elkader, which is about 15 miles from the Koehns' home, has an 18-hole golf course and several tennis courts that are open to the public.

Pat says vacationers are often surprised to learn that so many outdoor activities are available close by. For vacationers who are more interested in historic sites or shopping, Pat can recommend a number of places all within 25 miles of Little House.

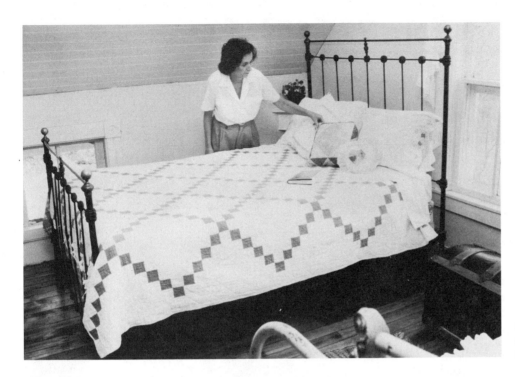

Cleaning and changing linens after guests leave is usually no big deal.

Little House is 30 minutes away from McGregor, Iowa, and Prairie du Chien, Wis., which have historic sites along the Mississippi River dating back to the 1830s and 1840s. You won't find large shopping malls in northeastern Iowa or southwestern Wisconsin, but small shopkeepers in these Mississippi River villages have local handcrafts, antiques and other reasonably priced items, according to Pat.

Scheduling Vacationers

Most of Pat's customers are families who want to reserve Little House for weekends or week-long vacations. Pat says there is rarely a problem with scheduling. Pat gets a lot of bookings during early autumn when the colors of the tree leaves are ready to peak. Usually, people call ahead of time and ask Pat when Little House is available.

"Little House seems to appeal to families most during the summer and early fall," she says. "The majority of our renters are also professionals from major metropolitan areas. They inevitably carry an armful of books to take advantage of the peace and quiet.

"We've had farm families from other areas of the state stay here while they visited relatives or friends. We've also had families meet at Little House for reunions. We generally have a group of college students rent the house in the winter for cross-country skiing. There is also a group of deer hunters who have been regular customers for many years during hunting season." No matter who rents the house, the Koehns have never had a problem with renters damaging or destroying property. People are also very careful to leave the house as clean as they found it upon arriving.

Pat and her daughter clean the house from top to bottom when people leave. This is seldom a big job. Muff keeps firewood available and handles all the minor repairs.

When Pat and Muff opened Little House, they knew it would not be a get-rich-quick business. The Koehns have paid off their initial investment, and Pat says Little House now provides the family with "fun money."

"This has been a good way for us to share a way of life we love with others," Pat says. "Rentals have been steadily improving and it does provide some income. But more importantly, we've made lots of new friends, which is equally important to our family."

(Interested in visiting Little House in the Woods? Write Patricia Koehn, Little House in the Woods, R.R. 1, Box 204-7, Elkader, Iowa 53043. Phone: (319) 783-7774.)

Reprinted with permission from Summer 1986 issue of *Rural Enterprise* magazine, P.O. Box 878, Menomonee Falls, Wis. 53051.

These Soybeans Earn Y25,600/Bu.

If you're not familiar with foreign currency, the soybean price in our headline is given in Japanese yen, and it translates roughly into $184 per bushel.

But to cash in on those kinds of earnings, you don't need to worry about international currency exchange rates. With a little initiative and marketing savvy, you may be able to command such premium prices by selling your customers on this ultimate seasonal taste sensation—fresh, green, edible soybeans.

"It's a great snack. After steaming them, people eat them like peanuts," says Tom Ney, director of food services at Rodale Press in Emmaus, Pa. "I think soybeans have the potential to become very popular in this form."

Ney has reason to be so enthusiastic. In Japan, people crave the crunchy, buttery taste of fresh, green soybeans so much they'll pay an estimated 250 yen (about $1.80) for 200 to 300 grams (about a half-pound), says Rick Davis, director of the Rainforest Information Centre in Kofu, Japan. "They are eaten everywhere—in homes, restaurants, bars and beer gardens," he reports. "They can certainly be called standard summer fare."

Any Variety Will Do

While "edible" soybeans generally mean tofu to most people, there are soybean varieties designed especially to be harvested green and eaten whole. ENVY and BUTTER BEANS are the best-known in the United States. The former is ready for harvest in 75 days, and the latter in a little more than 90 days, says Rob Johnston, president of Johnny's Selected Seeds (Foss Hill Road, Albion, Maine 04910), which sells edible soybean seed.

Both varieties will grow anywhere soybeans are traditionally grown—and even in cooler regions where they aren't. "We grow them here!" quips Johnston.

But if you're looking for a virtually risk-free way to test the idea, try harvesting a few bushels of conventional soybean varieties that you or a neighbor already have planted for use as fresh, green beans. Scientists at the Rodale Research Center (RRC) in Maxatawny, Pa., did just that a few years ago for an informal taste-comparison Ney conducted with edible varieties. "There didn't seem to be any significant taste difference between the two," Ney recalls.

There is a genetic difference between the two types, though. Edible beans are bred to be picked green, and their slightly higher oil content usually gives them a more buttery taste, says Skip Kauffman, RRC assistant director. Conventional soybeans are higher in carbohydrates, "so they tend to taste a little starchier," Kauffman observes.

Typical Japanese housewife enjoying a typical summer snack—fresh, green soybeans. In 1985, the Japanese consumed more than 80,000 metric tons of the crunchy, buttery-tasting beans, which retail for more than $3 a pound.

Cultural information is scarce. Kauffman suggests planting soybeans for fresh use at the same rate and time—and with the same equipment—that you'd use for any soybean crop.

Harvesting is the tricky part. Kauffman says there's a two-week harvest "window" that begins when the green pods have just started to fill out, and ends at the first sign of pod yellowing. In southeastern Pennsylvania, that period usually is during the last two weeks of August.

Once the pods have started turning yellow, it's too late to harvest the beans for fresh consumption, Kauffman warns. Carbohydrates have begun to accumulate and the beans will taste too starchy.

Johnston thinks large plots could be harvested by machines used for lima beans or peas. But both he and Kauffman hand-harvest the small amounts they grow. One obstacle to machine-harvesting is that the green, fibrous pods can be hard to separate from the branches. In fact, Kauffman often has to use pruning shears to clip the pods.

PYO customers can simply pull the entire plant. That's often how fresh, green soybeans are sold in Japan and Thailand, says Kauffman.

The beans must be washed thoroughly after harvest, then steamed for a few minutes to aid in protein digestion. Steaming also makes it easier to shell the beans. "It seems to soften the pod, and you can just sort of pop them right out," says Kauffman. The beans can also be served right in the pods, just like peanuts.

The lack of established markets for fresh, green soybeans in the United States doesn't mean Americans won't enjoy the product as much as people in other countries do, says Ney. "It's just that almost no one knows about them, here," he notes. Just think what a great "introductory" snack they'd make for your PYO customers.

Restaurants that specialize in trendy or exotic foods might be worth contacting, too, he adds. And whether you're selling to a family out for a Sunday shopping trip, or a chef seeking the latest in posh cuisine, be sure to have plenty of samples and serving suggestions on-hand.

Pricing is open-ended. Based on 56 pounds of soybeans per bushel and 139 yen to the dollar, Davis' retail-price estimate for fresh, green soybeans in Japan comes out to about $3.28 a pound, or nearly $184 a bushel. (And that's in a country where fresh, green soybeans are widely available.)

Of course, harvesting and preparing the crop is extremely labor-intensive, even on a small scale. But as the RRC's Kauffman puts it: "With the chance to earn $184 a bushel, I'd say it would be worth a try." That's especially true with a Clientele Membership Club and customers who will do nearly all the work.

The Rodale Test Kitchens have compiled a few recipes using fresh, green soybeans. For a copy, send $1 for postage and handling to: Soybean Recipes, The New Farm, *222 Main Street, Emmaus, Pa. 18098.*

Chapter 7
Marketing, The Most Important Part

From Pasture To Plate—Direct Marketing Of Livestock

Selling livestock directly to consumers is a more complicated process than selling pick-your-own sweet corn. Since you can't just put an awning up at the side of your feedlot and sell steers to passing motorists, you're going to have to become involved in processing—and abiding by the federal and state laws that govern the way animals are slaughtered. Direct marketing of livestock also requires careful planning, especially with beef, since production takes two years. And it requires effective salesmanship, because a side of beef, or even a naturally cured ham, is a big purchase for many consumers. Again, this is an area where your Clientele Membership Club (CMC) will prove invaluable.

Why do people buy meat from farmers? In a word, quality, according to a national study conducted by USDA. A Cornell University survey of 300 New York state freezer-beef customers confirmed that. Buyers said they wanted flavor, tenderness and juiciness first. Convenience and the availability of custom cuts were less important.

This emphasis on quality means that overfed animals and culls must still go to the packer. Meat sold directly to consumers needn't be graded, but anything below good quality discourages repeat sales.

There is also a growing market for meats produced without growth hormones or antibiotics. Perhaps one-tenth of the U.S. population has drug allergies, estimates Frederick W. Oehme, a toxicologist at Kansas State University's College of Veterinary Medicine. Many seek out organic farmers to supply them with meat. Michael Scully, an organic producer who owns a 400-acre grain, cattle and hog farm in Illinois, recalls the pleased response of his first customer, a truck driver who delivered his hogs to the stockyards. "He said he hadn't eaten pork for five years, because he got indigestion and nausea. He tried ours five nights in a row and didn't get sick."

Besides health-conscious consumers, who is most likely to buy from producers? You might expect your buyers to be city folks, but that's not always true. The Cornell study found that 67 percent of the freezer-beef consumers surveyed were rural or small-town residents. We know of one Nebraska farmer who developed quite a freezer-beef trade among neighboring farmers. Although they could no longer be bothered with slaughtering their own meat every fall, those farmers still wanted the top quality of homegrown beef.

Since buying a side of beef requires a big outlay of cash, it should be no surprise that freezer-beef customers are fairly affluent. Half of the 300 buyers surveyed in New York state had family incomes of $21,000 or more. A surprisingly large 88 percent of those consumers bought other foods directly from growers. The study also found that consumers seem to prefer smaller carcasses, which explains why smaller cattle breeds such as Angus are popular in the freezer-beef trade.

Without the guaranteed market a CMC provides, building sales volume is a major obstacle for farmers who sell livestock to consumers. Many of those farmers still sell most of their production at the sale barn, packing plant or stockyards, and at much lower prices. Illinois farmer Scully once tried to build a trade for his beef among restaurant buyers in Chicago and by advertising in consumer magazines, but was unsuccessful. Today, Scully sells about 15 percent of his annual beef production to individual consumers who pick up their orders at a custom packer in nearby Springfield. The rest of his cattle and most of the 350 fat hogs he sells each year are taken to the Springfield stockyards. John Hilgendorf of Welcome, Minn., has seen the demand for meat from his organic farm grow to the point where some of his customers drive to Welcome from Sioux Falls, S.D., and from Minneapolis, both about 100 miles away.

Yet, only about 10 percent of the 150 steers Hilgendorf sells annually are purchased directly by such consumers. He doesn't have a *guaranteed* market.

In the dairying country of New York state, where limited beef production and a scarcity of packing plants help keep beef prices low at conventional outlets, cattlemen have taken great pains to build direct sales. About one-third of the finished beef sold in that state goes directly to consumers. But most of those producers have a fairly small volume of business. "After selling to friends, neighbors and acquaintances, sales growth falls off rapidly at 30 to 40 head," says Cornell Agricultural Economist William Lesser. Again, lack of a guaranteed market really hurts.

One of the most successful New York freezer-beef producers is George Ambrose, who sells about 100 head annually from his part-time farm. His business hasn't reached that amount of sales overnight. Ambrose got his start during the meat-rationing days of World War II. When neighbors heard that he had meat available, they wanted some, too, and the business was launched. It expanded slowly to the 100-head level and has remained stagnant for some time, although Ambrose is constantly acquiring new customers as he loses some of his old ones.

Here's how Ambrose describes typical sales: Twice a year, in spring and fall, he writes his regular buyers to inform them of the availability of animals and the carcass-weight price he will honor for two months. At any other time, customers must contact him and ask for a new quotation. All sales are for sides, so Ambrose must be able to match two buyers before scheduling delivery. After notifying his customers when their meat will be ready, Ambrose delivers a steer to a local custom packer two weeks ahead of that date so that the meat can be properly aged. Buyers pick up their cut, wrapped and frozen meat from the packer. Ambrose always reminds customers to bring enough containers to carry their purchase. He also asks them to pay for their purchase when they receive their meat. A complete payment covers Ambrose's bill for the carcass, the packer's processing fees, and any taxes. Ambrose lets his established customers pay with a check; others must pay cash. Ambrose's cattle business has increased through word-of-mouth recommendations and he has used little advertising. Now that he is retiring from his regular job, Ambrose wants to increase his volume and is considering advertising at produce stands to attract customers used to buying direct from farmers. Promotion through newspaper classified advertisements hasn't proven very effective for New York producers.

Chuck and Shari Nycum of Benton County, Mo., are trying another approach to building a consumer market for all of the production from their 200 cows and 30 brood sows. They bought a small grocery store.

"We noticed that some freezer-meat customers cannot utilize a whole carcass at one time," Shari explains. "So we started selling halves and quarters, as well. A few years ago we bought a grocery store, and now we can use all types of beef and pork retail cuts. It works out beautifully. Some customers want more ground beef, for example, while others want more steaks and roasts. We now can accommodate more people with more product latitude." The Nycums hire a local locker plant to slaughter their cattle and hogs. They still serve a sizable home-freezer trade, but the retail store is doing more and more business. Their pork sausage, sold with three different spice mixes, is a hot seller. Business has grown despite a depressed cattle market during much of the start-up time.

The Nycums are sticklers about quality from production right through marketing. Chuck manages a straightbred Santa Gertrudis herd, from which he selects bulls to breed the Angus-Holstein crossbred cows that produce his beef calves.

"I don't put an animal on feed unless I have the meat sold on down the line," Chuck adds. "Many freezer-beef customers buy on a sort of layaway plan—so much per month while the steer is growing. I will even sell a calf still on the cow, let the customer pick out his own beef before the calf is weaned, and produce it to his specifications. We also feed enough cattle to provide the beef we sell at the store."

"We are farmers, not storekeepers," Chuck points out. "The store is a vehicle through which we sell the products of our farming operation. And we are in a rather unique location here, with the Lake of the Ozarks resort area close by. A lot of city people travel through here." His store, which carries a variety of food items in addition to meats, gives the ranching operation an outlet on a well-traveled highway. "It is paying off," says Chuck.

Retailing And Processing

Other farmers have made similar decisions to enter retailing or processing, and found that they gradually spent less and less time farming. Ted Carsten of Guilford, N.Y., spends much of his time marketing. The family business that he runs with his brother and father began as a livestock operation supplying a small direct-to-consumer trade. Today, it includes their 280-acre farm, a meat processing plant, a bakery and three health food stores in central New York state. The Carsten farm produces about 60 head of cattle and 50 hogs annually. Meat from their own herd and from another 50 hogs purchased from a neighboring farmer supplies the family's three Deer Valley stores and 55 other health food stores in New York, New Jersey and Connecticut.

Across the continent, Bob and Lee Markholt of Tacoma, Wash., have become increasingly involved in processing, while selling additive-free pork, poultry, beef and lamb to an urban clientele in the Seattle area. The business began as a small butcher shop for beef from the brothers' feedlot on their 9-acre farm near the Tacoma suburbs. In 1977, they began raising chickens without antibiotics, and experimenting with curing pork without nitrites. Two years later, they leased a $250,000 processing plant in nearby Puyallup to sell wholesale to about 25 health food stores and three supermarket-size food co-ops in Seattle. The Markholts still raise 15,000 fryers a year and 60 to 100 head of cattle, but to supply their expanding meat business, they have informal purchasing agreements with nearby hog and cattle producers.

Returns

Is direct marketing of livestock worth the effort? New York state producer George Ambrose thinks it is. He estimates that he can gross $30 more per head by selling freezer beef than if he sold the same animals through local auctions or direct packer sales. All of that isn't clear profit, of course, since extra time, telephone calls and postage are used to arrange individual sales to consumers. Although canceled orders and returned meat seldom occur, they must also be accepted as part of the costs of his business. To lessen the risk of bad checks, some producers accept credit cards, an arrangement that can be worked out with your banker for a small service charge.

A few direct-sales livestock producers, especially those who sell additive-free meat, ask a small premium over what they would get from conventional buyers. Lee Shrawder of Kempton, Pa., charges 5 percent more than current market prices for his organic beef. John Hilgendorf of Welcome, Minn., delivers his direct-sales livestock to a butcher 4-miles from his farm, and adds "2 or 3 cents over market price, because it takes a little extra time to handle one or two head."

"Freezer sales can be profitable. But, like any business, it requires additional attention and management, compared to selling through the easiest channel," cautions Cornell economist William Lesser. Having a CMC enables you to achieve the necessary high level of management—and profit.

He Bypasses 20 Middlemen

And controls his produce all the way from the fields to the consumers' dinner plates

---

NAPLES, N.Y.—When lettuce, tomatoes, sweet corn or other produce is harvested from the vegetable plots at Wild Winds Farm, it isn't packed into boxes, loaded onto trucks or frozen. It's cleaned up a little, made into meals and served at a restaurant on the premises. John McMath, owner of the farm, says he bypasses about 20 middlemen.

"I believe in the value-added principle," McMath says. "It can be well worth your time to process your produce in some way before putting it out in the marketplace."

His farm-restaurant is carving out a reputation for its country atmosphere, wholesome lunches and gourmet dinners,

Wild Winds owner John McMatch checks the sugaring house wood supply.

which are made as much as possible with fresh, organically grown ingredients. The restaurant's use of only fresh-picked herbs and its edible flower dishes have earned Wild Winds Farm a full-page spread in the *Washington Post* food section and a spot on the syndicated *PM Magazine* television show.

700 For Dinner

McMath, a former advertising executive, returned to New York's Finger Lakes region in the early '70s to shake what he calls "an addiction to sugar and alcohol." He and some friends used organic methods to restore the fertility of some worn-out cropland on his family's mostly wooded 500 acres. McMath now has a ruddy glow. Some of his soil has 11 percent organic matter. The restaurant once served strictly vegetarian fare, and attracted only a trickle of customers. Now, "Trout a la Maison," and "Tournedos Henri IV" are on the dinner menu. "We had 700 dinners in one day last season," McMath says.

The quality and presentation of Wild Winds' entrees prompted one newspaper food editor to describe the restaurant as a "must" stop in the Finger Lakes region. So many tourists and locals *are* stopping that the business is growing too quickly, McMath says. He is looking for ways to increase the parking area without disrupting the farm. Twenty people work full time at the farm and restaurant during the summer season's peak. The business provides enough income to support McMath's environmental education efforts, which include guided walks on 3 miles of nature trails through the farm's pasture and woodlands. He built an old-fashioned maple sugaring house complete with a wood-fired evaporator. Only 50 trees are tapped each spring with the help of school groups interested in learning how to make maple syrup.

But it's the meals, prepared with a steady flow of homegrown produce, herbs, edible flowers and small fruits from 2 intensively cultured acres, that keep customers coming back for more. "A lot of people still think of us as 'that crazy health-food place,' but our image is changing," McMath says.

Freshness First

Ellie Clapp, associate director of the farm, says, "This food can't possibly be any fresher." Such variety and quality is just not available from wholesalers at any price, she adds. But matching the supply with the demand can be tricky. Planning for the growing season goes on through the winter.

A light lunch in the country store cafeteria.

"Each year we get a better idea of how much we'll need, but we always try to overplant a little," Clapp says. This is crucial with the edible flowers, which are the restaurant's specialty. "When people come here they expect to sample edible flowers and have them presented with the meal. If you don't come through, word will soon get around that your quality is slipping."

The real work begins in early spring when seeds are sown in flats and set out in the farm's solar-wood-propane-heated greenhouse. All plants are grown from seed, saving the expense of buying from a commercial greenhouse. Many crops are direct-seeded as soon as the ground can be worked. As lettuce, spinach, over-wintered parsley and other crops start coming in, so do the customers.

The edible flowers — nasturtiums, marigolds, violets and others — get special attention. Many are grown in shaped beds that are both ornamental and functional. One tiered bed is visible from the porch dining area. Other beds are in or near the vegetable plots and can be seen from the main dining area. Other edible flowers, such as day lilies and dandelions, are foraged from the surrounding fields and stream banks. Most first-time customers don't know much about edible flowers. Clapp says they are often pleasantly surprised to see familiar "faces" like squash blossoms or rose petals perched on a dish of butter or garnishing a soup.

When the plots are in full production, spot insect outbreaks are treated with rotenone. But McMath says he doesn't have the insect and disease problems experts say he is supposed to. McMath thinks healthy soil and healthy plants are the best insurance against pests and diseases. He rotates vegetable crops each year, sometimes with a cover crop, and avoids planting large blocks of one type of vegetable. He says monocropping allows minor problems to develop into disasters. Compost is added to the soil as a mulch and between plantings. A five-stage composting area is maintained next to the plots.

Vegetables and small fruits are harvested the same day, sometimes even within the same hour they are served. In fact, on some busy days, diners might very well see their own salad being harvested as they look out of the large dining-area windows, Clapp says. The harvest-as-you-go method can be hard on the nerves, she says, but it is a great advantage with items like red raspberries, which are often served with champagne; and sweet corn, which is best when it's in the pot just after it's picked.

Parsley, chives, thyme, sage, tarragon, savory, dill, many varieties of mint and other herbs are grown — herbs which, if bought fresh, would cost dearly.

Eggs are purchased in bulk from a nearby farm. Bread and nitrite-free meats produced under the Wild Winds label are served as well.

It takes a good location, some business savvy and a reputation for quality to build a successful restaurant-farm, McMath says, but it can be a profitable alternative to conventional markets.

Target New Markets

Once the farmer's foes, animal rights and environmental groups are becoming your new partners for profit.

Remember when "Sugar Frosted Flakes" became "Frosted Flakes"? Or when the word "Natural" started showing up on packages in big supermarkets, instead of just the tiny health food store across town?

At first, farmers had little reason to be interested in these subtle changes. But then, public concerns about the health value of food began affecting them more directly. Doctors told people to eat more fresh fruits and vegetables, and less red meat. Newspapers warned of groundwater pollution by farm chemicals, and antibiotic residues in meats. The government banned or restricted popular pesticides like Lasso, EDB and Temik. Grocery stores began refusing shipment of apples treated with Alar.

And before long, America's food-buying habits had become the object of a multibillion-dollar tug of war. "It's an issue that won't go away, and anyone who thinks otherwise is like an ostrich, head buried in the sand and vulnerable to a fatal attack," wrote *Farm Chemicals* Editor Charlotte Sine in a 1986 editorial.

Sine is right. More and more people want assurances that the food they buy will not only satisfy their appetite, but "increase fitness and energy, and add years to their lives," as the Food Marketing Institute (FMI) put it in a 1986 survey. And to Sine's readers—the people who make and sell agricultural chemicals—that's probably not very good news.

But it's great news to farmers. Because now, reducing chemical use isn't just a good way for them to save money, it's a good way for them to make money, too. How? By cashing in on the many new markets emerging from these overlapping trends, which include:

- Fast-food franchises.
- Upscale restaurants.
- Grocery stores.

Fast Food To Fine Dining

"I think there will be an expanding market for organic food. And I also feel there will be opportunities somewhere between spraying everything you have at one end of the spectrum, and organic at the other," says Michael Jacobson, executive director of the Washington, D.C.-based Center for Science in the Public Interest (CSPI).

CSPI is one of several groups that are working directly with farmers to expand the sale of organic and/or locally grown foods all across the country. It's an informal movement, to be sure. But its participants are varied enough to accommodate virtually everyone. For example, at four fast-food restaurants in Maine, customers can load up their salad plates with organic vegetables produced on nearby farms. Meantime, upscale diners in New York or Chicago can savor the taste of free-range veal raised under the strict supervision of—are you ready for this?—an animal rights group. And Colorado shoppers can select from sprayed, unsprayed or totally organic products at supermarkets in Denver and Boulder.

To Jacobson, these developments illustrate an equally subtle approach to promoting healthy food, yet one that's more helpful to farmers than just changing words on a package. "Instead of being *against* drugs and pesticides, (we) will be *in favor of* safe food," he explains.

In fall 1986, CSPI launched a nationwide campaign to promote that view. Known as "Americans For Safe Food," the program features a multimedia blitz that includes pamphlets, radio and TV appearances, and magazine ads, all carrying one simple message: When it comes to many foods, "What you see is not what you get."

The whole idea is *not* to give farm chemicals a bad name, Jacobson stresses. In fact, he'll be the first to admit that eliminating chemicals is an option not all farmers have. "But that doesn't mean people shouldn't have the opportunity to buy food grown without them, at least at certain times of the year," he says.

Jacobson hopes to give consumers that opportunity by compiling a national directory of farms and other places from which they can mail-order food grown with few or no synthetic chemicals. (Like to be on that list? Write: Deborah Schechter, Americans for Safe Food, 1501 16th Street N.W., Washington, D.C. 20036. Phone: (202) 332-9110.) He's also encouraging people to demand that supermarkets buy more local and organic food.

Jacobson realizes the program's success hinges on whether consumers will pay the higher prices many organic foods command. But he cites numerous reasons for optimism: There's the Grand Union supermarket chain, whose 400 stores currently sell "Natural Beef" from the Coleman Ranch in Colorado "at a premium price quite successfully." There's entrepreneur Stan Rhodes of Oakland, Calif., who has enrolled dozens of farmers and retailers in his "NutriClean" program for labeling the nutritional and pesticide residue content of produce sold in grocery stores.

And finally, there's the much-publicized Food Marketing Institute survey, which found that chemical residues in food are a major concern to three-fourths of all shoppers.

Fewer Sprays 'Make Sense'

Still, many people feel Jacobson overestimates the public's fear of pesticides, and its willingness to spend more for food grown without them. "I think people presume food in the supermarket to be safe," says consumer researcher Chip Certain of Castle & Cooke, owners of the Dole label and one of the world's largest food companies. "I don't think you're going to find that people are willing to pay more for (food grown without chemicals)."

Certain discounts the FMI survey, because the participants were asked to rank a pre-listed group of "concerns," rather than listing their own. Instead, he bases his views on a series of interviews he conducted in spring '86 with people who'd recently bought fresh produce.

Asked if they'd pay more for products guaranteed to contain no harmful chemical residues, "The reaction was, 'Hell, no!' They felt it was the government's job to be sure the food was safe." In fact, Certain says many of the people suspected grocery clerks might neglect to remove "organic" signs after such products were sold out and replaced with sprayed ones.

Certain conducted the study in part to determine whether Castle & Cooke would benefit from having its fruits and vegetables tested with the NutriClean system. "Using less pesticides is something we would favor," he explains. "It costs less, and anything that reduces the cost of growing food makes sense to us."

But in Castle & Cooke's case, the investment simply wouldn't pay off, he says. Too much of the firm's income is from imported products which must be sprayed to avoid exotic insect and disease outbreaks. Even if they could find non-chemical ways to reduce these risks, the added cost of testing the products would force a price-hike that—according to his findings—customers wouldn't tolerate.

To Certain, it was a somewhat disappointing result. "Stan (Rhodes of NutriClean) will tell you that what they do will only reward good agricultural practices," he says. "And quite frankly, he's right. Good agricultural practices lead to less use of insecticides and herbicides."

That last point is sure to be the subject of debate for many years. But consumer advocate Jacobson sees no reason to make farmers who use fewer chemicals—and shoppers who want to buy their products—wait for the final outcome. "Many consumers are saying, 'I don't want to worry about . . . who's right. I just want uncontaminated food.' We're saying, give people that alternative."

And several other groups agree.

Animal Activists Boost Meat Sales

CHICAGO—Most veal or egg producers would take one look at Bob Brown's business card, see that he's founder and president of the Food Animal Concerns Trust (FACT), and toss it in the wastebasket.

Outdoor veal pens are a rare sight on most veal operations. But FACT, for whom these calves are being raised, is a pretty rare group of animal activists.

But for six farmers in the Northeast and Midwest, such a move would have cost more than a quarter million dollars in sales each year. Because, unlike many animal rights advocates, Brown doesn't use pressure tactics to promote his cause. Instead, he seeks out farmers willing to use "humane" livestock production methods. "Then we make sure it's profitable for them to produce according to our standards," he says.

The idea began in 1982, when Brown formed FACT "to do something about factory farming. We pretty much concluded that media campaigns wouldn't work, so we thought the way to do it would be to develop some alternative products, then go directly to consumers."

Rambling Rose veal is the group's newest and possibly most unique product. Two farmers near Stuyvesant, N.Y., produce a total of four calves a week under FACT's standards, which include weaning them at their own pace, allowing them to roam freely in outdoor pens, and feeding drugs or antibiotics only when absolutely necessary.

The meat, which is sold wholesale to upscale restaurants in New York and Chicago, earns about $84,000 a year. But more importantly, Brown ensures the farmers a $75-per-calf profit, which is far more than they could otherwise earn on such a small scale.

Even more popular are FACT's Nest Eggs, which are produced from uncaged layers and sold in some 50 markets throughout New York, New Jersey and Chicago. Three New Jersey farmers and one in Illinois ship a total of 2,200 dozen eggs a week, with a retail value of $225,000 a year. And like the veal producers, they need only follow FACT's guidelines to make money. "I think it's a good set-up," says Jim Abma, who raises 7,000 layers on his 30-acre farm near Wyckoff, N.J. "We've got a set price, we know exactly what we're going to make, and it's profitable."

Brown pays Nest Egg producers 88 cents a dozen for unprocessed eggs. "If we supply the pullets, then he'd get about 18 cents less than that," he says. "And if he processed them (washed, graded and packed), then he'd get an extra 9 or 10 cents."

As an established egg producer who has never used cages, Abma already had the necessary skills and equipment. FACT bought the 1,500 Rhode Island Reds that would lay the eggs. "We just had to give the birds a little more room (2.75 square feet each, compared with 2 square feet in his other buildings). It really wasn't much different from what we'd been doing," he adds. Abma also had to switch to FACT's recommended feed, and eliminate debeaking.

Abma feels the system is perfect for small-scale egg producers. "Big ones rely on square feet per bird; they can mechanize and spend less," he explains. "But for a small farmer in a rural area who doesn't have other opportunities, it's certainly worth looking into."

Brown estimates an experienced farmer with 3,000 hens would earn roughly $10,000 profit a year by producing Nest Eggs. And he's hoping more farmers will take advantage of the opportunity. "We're interested in (adding) new farms around big cities like New York, San Francisco or Los Angeles," he says. "Animal rights—to small farming—could be a tremendous boon. There's no question that these products are good to eat. That's why people buy them." Write: FACT, P.O. Box 14599, Chicago, Ill. 60614. Phone: (312) 525-4952.

Speciality Sales Top $16 Million

BOULDER, Colo.—"There are plenty of people around here going broke raising wheat," says truck farmer Chet Anderson. "But for good, fresh, organically grown produce, there's a great market."

Anderson should know. He's one of several Colorado farmers earning premiums of 20 percent or more by growing organic and "pesticide-free" produce for Alfalfa's markets in Denver and Boulder, whose combined yearly sales have soared to $16 million since 1980.

While Alfalfa's exists mainly to serve health-conscious shoppers, it also gives farmers like Anderson a chance to expand their crop mixes and markets, says Greg Kurtz, produce manager at the Denver store. "The availability of locally grown, organic produce has increased 400 to 500 percent over the past five years," he says. "Farmers around here know where to go to sell it."

What makes Alfalfa's unique is that, unlike many supermarkets (and even some health food stores), it doesn't require farmers to meet rigid sales terms.

That often means Alfalfa's has to rely on out-of-state sources for products, such as the poultry it buys from an organic farm in Arkansas. Likewise, if organic produce isn't available, Kurtz must substitute non-organic crops (and identify them with signs in the store), or try convincing a farmer to eliminate sprays, like he did a few years ago with a tomato grower in Boulder. "We told him, 'We'll pay you more if you don't use chemicals. If the market price is 30 cents (a pound), we'll pay you 45 cents.'" The farmer didn't hesitate for a minute.

Such informal arrangements force Alfalfa's to vary the prices it pays growers. Kurtz might pay 15 cents a pound for non-organic zucchini (about the same as any grocery store), and 25 cents for a pesticide-free crop grown with synthetic fertilizer. In 1986, he paid 30 cents a pound for totally organic zucchini.

Health-food purists might consider Alfalfa's flexibility unacceptable. But Kurtz feels he helps more farmers—and customers—by simply selling the best local products available. "A lot of our farmers are in between," he says, explaining that some use chemicals when necessary, while others—like Anderson—avoid them at all costs. "That's why we make distinction between organic and pesticide-free."

And to Anderson, who sells one-third of his produce and herbs to 25 Denver-area restaurants, that's good news. "By and large, the store pays less than restaurants," he says. "But you can't sell 75 cases of lettuce a week to a restaurant. (Alfalfa's) gives me a market, and a good, fair price."

Fast-Food Chain Buys Local, Organic

SOUTH PORTLAND, Maine—Fast food is sprucing up its image in this town along Maine's southern coast. And in the process, at least four local farmers have found a steady, profitable market for organically grown fruits and vegetables. "Since 1979, we have been buying organic produce from local farms for our salad bar—just about everything," says Ken Raffel, co-owner with his brother of four Arby's Roast Beef franchises, the nation's seventh largest fast-food chain.

Though Raffel never intended to limit the program to organic farmers, "those who have been most interested in getting involved have been organic," he says. That is fine with Raffel and his customers. "It's just such wonderful produce. It's the most delicious we get all year."

One of Raffel's biggest suppliers is Jim Economou, who grows 40 acres of fruits and vegetables in Woolwich. "They are our largest wholesale account," says Economou, noting that the $5,000 worth of produce Arby's buys from him annually is more than double what his next largest restaurant customer buys.

Though Economou sells much of his produce at his own retail stand about 8 miles away from the farm, he relies on restaurants for nearly 60 percent of his income. The reason: He can sell a larger volume at higher prices than grocery stores pay. For example, he sells tomatoes for 75 cents a pound at the retail stand, and 60 cents a pound to Arby's. The restaurant also pays 25 cents a pound for cabbage, "which is pretty darn close to retail," he notes.

"We're not cheaper or more convenient," Economou adds. "What we have to offer is freshness and quality." Raffel appreciates that. "We may pay a little more, but we get a much better yield," he says. "If we buy broccoli from California, for example, by the time it gets here we usually get a lot that's not usable."

Apparently, other local restaurants feel the same way. "I think there's a lot more potential for this than farmers realize," says Economou. "Nine out of 10 places we approach are interested, yet they've never been approached before."

One exception is the other local fast-food restaurants he's visited, all of which have turned him away. "We've sold to restaurants for eight years, but have never been able to sell to fast-food places until Arby's," he recalls. "They have set themselves apart in a good portion of the public's eye up here. And that's been a breakthrough for farmers and consumers."

Soon, Raffel hopes to separate his Arby's franchises even further from the pack. He's currently exploring ways to sell organic beef from nearby Wolfe's Neck Farm, a 300-head operation run by the University of Maine.

Many details remain to be worked out, but Raffel feels the best approach would be to feature "organic hamburgers"—and possibly even retail cuts of meat—on special promotional days.

Why? "To become better integrated into the local economy, support local farmers and have the best quality products for our customers," declares Raffel.

'Billboard' In The Sky Boosts Sales 250%

MOSINEE, Wis.—When apple-grower Ken Ulrich began looking for new ways to lure retail customers to his farm, a billboard seemed like the perfect solution. Problem was, a local ordinance prohibited him from erecting such a structure on his property.

Today, it's not local zoning laws that govern Ulrich's promotional efforts. It's Part 101 of the Federal Aviation Administration Regulations—a section devoted to "Moored Balloons, Kites, Unmanned Rockets and Unmanned Free Balloons."

That's because the 'billboard' Ulrich finally put up isn't really a billboard, at all. It's an 11- by 13-foot, helium-filled balloon shaped like an apple.

"The results were fantastic," says Ulrich. "In 1986, the first year of using the balloon, our sales went up by 250 percent. Demand was so good that we not only quit advertising in the media, we also raised our prices from 39 cents to 55 cents per pound for our apples. It really turned things around for us."

Ulrich is no stranger to farming or the friendly skies. He grew up on a Minnesota farm, and is currently a pilot for a major commercial airline. In the late '70s, a company merger resulted in Ulrich being transferred from Louisiana to Wisconsin, and he saw the opportunity to put some of his farming experience to work.

A major obstacle was his erratic schedule, which often keeps him away from home for days at a time. "This made livestock farming impossible," Ulrich explains. "I just could not be tied down with daily feeding chores."

The logical alternative was retail fruit sales, which would let him take advantage of the farm's unique location and built-in labor force. "Because we are located near a four-lane highway, I thought we had lots of potential sales from tourists traveling through," he says. "Plus, my family and I could manage these crops while I kept my off-farm job."

Drive-In Profits

Ulrich chose apples as his main crop for his 40-acre farm, and raspberries and honey as sidelines. But finding a unique promotional strategy became difficult. He was spending roughly $500 a year for newspaper ads, alone, and occasionally ran spots on local radio and TV stations. "Radio, TV and newspaper advertisements just eat you up," says Ulrich. "We had to find a more cost-effective method of marketing."

What he came up with was an idea that made use of one of the oldest marketing gimmicks known: Create a unique attraction that will lure customers to the farm. *Then* worry about making the sale.

From regular flights to Northern Plains, Ulrich learned of a Sioux Falls, S.D., company named Aerostar International Inc., which has designed and built inflatable promotions for such companies as Coca-Cola, Kodak and Macy's department stores. He quickly realized the impact one of the firm's balloons would have on both local residents and highway travelers. "The balloon would not only advertise the business, but would make potential customers curious enough to pull off the road to take a closer look," he explains. "Once we get them into the store, we can almost always make a sale."

Ulrich contacted Aerostar International, and soon was the proud owner of a 13-foot-high apple balloon.

Made of a tough, nylon fabric and anchored by a 150-foot-long steel tether line, the balloon is visible within a three-mile radius of Ulrich's farm. That gives the structure many advantages over conventional advertising, Ulrich says. "First of all is the flexibility. With a sign, you pay for 12 months, even though you aren't marketing all through the year. With the balloon, you take it down and put it away at the end of your marketing year.

"Also, you can take this with you to fairs, shows or wherever you want to set up a temporary booth," he adds. "It is a real attention-getter."

A more subtle benefit is that, after highway travelers spot the balloon, they have plenty of time to decide whether to stop and shop. "When you're using a sign, a driver has very little time when traveling 60 miles per hour to not only read the sign, but also to make a decision," explains Ulrich. "The balloon gives them time to think about their choices longer."

The type of customers attracted by the balloon is yet another advantage. Local print and broadcast ads drew time-consuming phone calls from people in Ulrich's immediate area. Often, they were just window-shopping for the best prices. The balloon attracts a good many tourists, who don't have time to compare prices, Ulrich says.

Fruit Fills The Air

Besides a separate listing for the orchard in his phone book, the balloon is now Ulrich's only 'ad.' In fact, Ulrich was so impressed with the results of his balloon that he has formed his own company, High Sign, Inc., to sell helium-filled horticrops to other farmers.

Ulrich takes orders from interested growers, and contracts with the South Dakota firm to design and manufacture them. "Besides the apple, we now sell a strawberry and are in the process of coming up with pumpkins, peaches, and raspberries," he notes. "We're even working on a Christmas tree."

MARKETING, THE MOST IMPORTANT PART

After about six months in business, High Sign Inc. had sold six balloons and was receiving about 15 inquiries a week.

The balloons retail for about $2,000 apiece, and are easy to patch if torn (a repair kit is included). Ulrich says they require about $150 worth of maintenance and helium a year, and the average life expectancy of the unit is about eight years.

"One of the questions I get most often is about windy days," he notes. "I've found that, when the wind gets to about 25 or 30 mph, the balloon should come down." Although it's unlikely the tether line would break, if the balloon were to get away, it would deflate at higher altitudes and descend who knows where, he explains.

Besides his small, on-farm retail market, Ulrich is constantly seeking new ways to sell his products. He is now working with a local cheese factory to assemble gift packages containing cheese, honey, maple syrup and apples. "There has been a good deal of growth in these kind of gifts the last few years," says Ulrich. "They are perfect for the person who has everything."

The gift packs will be marketed in Ulrich's retail store, and in a retail outlet at the cheese factory. The factory also will distribute them to some 40 grocery stores and other retail outlets. Each gift pack will display the "Something Special From Wisconsin" label, which the Wisconsin Department of Agriculture is encouraging growers to use to identify locally grown products.

Ulrich also has begun seeking U-pick customers for his 2 1/2 acres of raspberries—a project handled mainly by his three children still living at home. "They take care of the plants and direct the people who come out to pick," says Ulrich. "It's good for them, because it gives them experience and puts a little money in their pockets, as well."

Year-Round Sales

Soon, Ulrich hopes to start a U-pick strawberry patch to complement his apples and raspberries. "Then we would have produce to sell throughout the growing season: Strawberries in June, raspberries in July, and apples in August through December," he says. "With this plan, we would make better use of our labor and facilities."

Ulrich grows 23 different varieties of apples, giving him a mix of colors, types, and ripening dates. All of his trees are dwarf types on M26 and M7A rootstocks. The former is winter-hardy, but is more susceptible to drought and fire blight. The latter is less winter-hardy, but offers good blight-resistance and a long taproot to draw moisture from deep in the soil during periods of low rainfall.

He also grows several varieties of raspberries, sells farm-produced honey in his retail store, and rents his hives to local cranberry growers for pollination.

One of the keys to good marketing is that consumers have a healthy image of your product, stresses Ulrich. He has committed himself to using as few pesticides as possible in his orchard. "We use about one-third the amount of spray

"Mission control to Applestar... All systems go." When flown 150 feet in the air, Ulrich's apple balloon is visible for three miles in all directions. Apple sales increased 250 percent the first year he used the device.

we once did," he says. "By monitoring weather, we can spray only when needed. Also, we mow between the trees, rather than using an herbicide. It helps not only our image, but our bottom line, as well."

But to many orchardists in central Wisconsin, insects and weeds are minor problems compared with white-tailed deer, which cause heavy damage by feeding on the fruit and shoots of young and mature trees, alike.

The problem is so severe that the Wisconsin Agriculture Department annually provides millions of dollars to farmers for fencing and other deer controls. To be eligible for the state funds, a farmer must prove that crop losses to deer damage are at least $500 per acre.

Ulrich has never worried about qualifying for such a grant. He's come up with a much less expensive solution. "We use two means of control: We have a good watchdog, so we put his doghouse out in the orchard. Second, we get a nearby motel to give us all their used soap, which we put through a sausage-maker to make soap pellets. We spread these pellets throughout the orchard.

"I don't know if it is the scent of humans or just the soap, itself, but I do know it has made a real difference in the amount of (deer) damage we've had since using it."

The result of all this is a farm where diversity is practiced at all levels, from crops and production methods to advertising and marketing. As Ulrich puts it: "I'm a firm believer in hard work and not limiting yourself when working toward a goal."

Editor's Note: *For more information on the balloons, write: High Sign Inc., 1486 CTH DB, Mosinee, Wis. 54455. Phone: (715) 693-6201.*

Manna From The Mail

It takes careful planning and decent up-front money, but direct mail selling can help you tap profitable new markets.

Whatley-style, diversified farms lend themselves to a thriving mail order business. Many of the items produced on these farms are naturals for the mail order trade. They include: sweet potatoes, nuts, kiwifruit, guinea fowl, game birds, honeybee products, quail livers, rabbits, fresh-pressed grape juice and even venison.

The two factors that have contributed most to the boom in selling food by mail are improved packaging and overnight delivery service to virtually anywhere in the country. For example, a salmon can now be swimming in the Pacific Ocean, 50 miles offshore from Seattle, Wash., at noon on Monday. By noon the very next day, that same fish can be on my kitchen table in Montgomery, Ala.

The Name Game

To sell successfully by mail, the one thing you absolutely, positively must have is a potential customer's name and address. Or, more correctly, a whole list of the names and addresses of potential customers.

Where can a farmer get these names? You could pull them out of the local telephone directory at random, but your mailing would almost certainly be a dismal failure. If you're lucky, you might be able to pay for the cost of the postage stamps, envelopes and photocopies that went into the mailing. That's because you just broke all the rules of direct mail marketing. It's not enough to have any old list of names. For a direct mail offer to be truly successful, you need the right list, a list that will give you the best possible prospects for a good response—of paid orders. *That calls for finding a list of people who you know ahead of time are interested in what you have to sell.* Of course, the very best list to start with is right at your fingertips. It is your Clientele Membership Club roster.

If you're still starting up your CMC or want to reach a much larger audience than your 1,000 or so members, your best bet is to rent a mailing list from a reputable list broker. That's right, a list broker. There really are people who make their livings by renting mailing lists from magazines, book clubs, department stores, credit card companies, catalog merchandisers and the like. After all, who do you think is responsible for all the "junk mail" that constantly shows up in your mailbox? List brokers, that's who. (There are also list managers and even list compilers, too, but you won't have many dealings with them.)

Selling directly through the mail got its formal start in 1872 with the introduction of Montgomery Ward's first catalog. But it didn't really take off until the middle of the next century, when high-powered computers and zip codes suddenly made it possible for direct marketers to reach into the average American home—and pocketbook—with the precision of a brain surgeon. Direct mail marketing now generates some 10 billion pieces of mail a year, at a cost of more than $2.5 billion, according to the Direct Marketing Association Inc. (DMA) in New York City. The DMA is made up of more than 2,000 direct marketing companies (from 35 countries around the world) that earn more than $150 billion a year by selling everything from books and magazines to gold coins, insurance, credit cards and slinky lingerie directly to consumers. They're doing exactly what I tell farmers they need to do—*eliminate all middlemen.* Sell directly to consumers.

It takes big bucks to join that crowd, though. It can cost up to $18,000 a year to join DMA (membership dues are assessed based on a firm's expenditures or sales). But you don't have to belong to the association to benefit from its considerable direct marketing brain trust. Each year, DMA publishes a book that lists companies specializing in helping direct marketers find just the right mailing lists. It's called the "Service Directory for List Brokers, Managers and Compilers." Cost is $30 for nonmembers. In addition to brokerage houses and contact names, addresses and telephone numbers, it includes brief but vital descriptions of each broker's area of expertise. Look at the broker's track record, with special emphasis on his experience in the market you're trying to reach. After all, you wouldn't want a broker who had only handled lists for people selling lingerie if you want to sell country smoked hams. Write:

Direct Marketing Association Inc.
Publications Dept.
6 E. 43rd St.
12th Floor
New York, N.Y. 10017

Get 'The Right' Lists

How important is the mailing list? Here's how the DMA puts it in a special report on the subject: "A mediocre mailing package sent to an excellent list will generally prove more profitable than an excellent mailing package sent to a mediocre list. And the combination of a well-executed, appropriate package with a strong offer going to a carefully selected list still proves to be unbeatable in terms of pulling in orders and responses."

How do you know which lists to try? Find lists by using the lowest common denominator. In this case, that means an overriding interest in good food, coupled with a proven responsiveness to direct mail marketing. If names like "Harry & David's," "Figi's," "Wisconsin Cheeseman," "Neiman-Marcus" or "Swiss Colony" sound familiar, you already have some ideas for lists that might work—and all just from looking at the promotional mailings in your mailbox. Or maybe you're thinking about up-scale epicurian magazines like *Bon Appetit*, *Food & Wine* or *Gourmet*.

Don't forget publications about growing food. One such magazine that generated hundreds of new subscriptions for my *Small Farm Technical Newsletter* was *Rodale's Organic Gardening*. How about publications that cater to a country lifestyle? Think about magazines like *Southern Living*, *Yankee*, *Mother Earth News*, *Harrowsmith* (the U.S. edition only, you don't want the hassle of mailing promotions and products into Canada), *Farm and Ranch Living*, *Country Journal* and *Country Living*.

What source of names guarantees that every single person on the list is seriously interested in food and spends a goodly amount of time in the kitchen? Why lists of cookbook buyers, of course.

Bounce those ideas off of a list broker, for starters. Brokers generally know which lists or list segments work for which offers and products. Unless a list is very small, you'll more than likely end up renting only a portion of the list. In most cases, the minimum order is 5,000 names, which is what the experts say you need to accurately test a new list. Don't let that figure of 5,000 go to your head, though. The fact that you're mailing to 5,000 people does not mean that all 5,000 of them are going to buy your smoked hams or whatever you're selling for $39.95 each. So don't multiply 5,000 times $39.95, come up with the staggering sum of $199,750 and book a seat on the first flight to Rio.

Not even one out of every 10 of those people will buy a ham from you. Your returns are more apt to be something like two or three out of every 100. In direct mail marketing, a response of 2 percent to 3 percent is generally considered fairly respectable. Of course, response rates can—and do—go much higher. They can also go a whole lot lower, right on down to an absolute zero if you send the wrong offer to the wrong list. Careful list selection, as the Direct Marketing Association points out, is your best guarantee of success. The right lists will make a mediocre offer shine and a good offer positively radiant.

Takes Money To Make Money

Nowhere is that old adage truer than in the direct marketing business. Remember all of that "junk" mail cluttering up your mailbox? Now, let's translate that into dollars and cents. It's not at all uncommon for those promotional packages to cost 30 cents apiece. Now, if 30 cents doesn't sound like much, quickly multiply that by 1,000. Three dimes suddenly become $300. That's the amount a lot of direct marketers are now spending to put 1,000 of their promotional packages in the mail. Those using a lot of 4-color printing, slick paper, big catalogs and original artwork are spending more, much more.

Your promotional packages will be nowhere near that sophisticated or expensive, at least in the beginning. For starters, your package may consist only of an outside mailing envelope and a photocopied letter with an order form.

Using a postage-paid return envelope (called a BRE—business reply envelope—in the trade) helps increase response. The whole idea behind BREs and other response aids is to make it as easy as possible for the customer to respond to your offer. While you can mail without such devices to minimize costs, the one cost you can't avoid is that of names. And names aren't cheap. For example, if you want to mail an offer to readers of *Bon Appetit*, it will cost you $75 per thousand, just for the names. The minimum order is 5,000 names (5 x $75 = $375). Then add $3 per thousand for selection by state. Other "selects" are available by magazine renewal rates, sex, county population, zip code, and myriad other factors, all at additional cost, of course.

Peel-off labels to go on your mailing envelopes cost another $5 per thousand. Shipping fee is $10. It's kind of like buying a new car. By the time you get done adding options, the base price doesn't mean a whole lot.

On first-time orders or when dealing with small-businessmen like farmers, brokerage firms require that orders be prepaid. Orders should include the number of names you want, your mailing date and a sample of your package. Most list owners won't rent names without seeing your package, first. They want to see what you're sending their customers. In addition to that, they also check up on you by "seeding" their mailing lists with fictitious names at their business or home addresses. You can't blame them for that. After all, they just want to protect their good name and maintain the goodwill of their customers.

The broker's marketing and list-selection advice is "free." When you actually place an order for names, the broker will usually keep 20 percent of the cost of the names you're renting and forward the rest to the list owner or the owner's list-management firm.

The following Whatley Farms' sweet potato mail order offer is an example of the simple yet effective approach you can use for any item you're trying to sell by mail.

Dear John and Mary Shackelford:
(use your personal computer and Clientele Membership Club list to personalize the salutation like this)

SWEET POTATO THANKSGIVING AND CHRISTMAS PACKAGES

Whatley Farms Inc. will have 40-pound packages of cured U.S. #1 "Carver" and "Jewel" sweet potatoes available for the Thanksgiving and Christmas holidays, beginning Oct. 20.

These sweet potatoes, which I developed at Tuskegee Institute, are far superior in appearance, freshness, taste and quality to any sweet potato you will see in your local supermarket. They would make an excellent addition to your holiday dinners. And, they also make excellent gifts.

These packages will sell for $22.50, plus United Parcel Service (UPS) postage and a $1 handling fee. Just take the first three digits of your zip code and use the following chart to determine the correct UPS postage.

(UPS Chart)

	Zip Code* Prefixes	UPS Zone	Price	UPS Postage	Handling Fee	Total
1.	010-016	5	$22.50	$ 9.39	$1.00	$32.89
2.	150-156	4	22.50	7.63	1.00	31.13
3.	330-333	5	22.50	9.39	1.00	32.89
4.	350-362	2	22.50	4.77	1.00	28.27
5.	376-383	3	22.50	6.08	1.00	29.58
6.	600-628	4	22.50	7.63	1.00	31.13
7.	747-762	4	22.50	7.63	1.00	31.13
8.	800-831	6	22.50	11.83	1.00	35.33
9.	840-844	7	22.50	14.43	1.00	37.93
10.	846-853	7	22.50	14.43	1.00	37.93
11.	900-928	7	22.50	14.43	1.00	37.93
12.	936-961	8	22.50	17.37	1.00	40.87
13.	980-986	8	22.50	17.37	1.00	40.87
14.	988-993	8	22.50	17.37	1.00	40.87

When your package arrives, you should immediately wash and grease each root with bacon drippings or vegetable oil and BAKE IN A CONVENTIONAL OVEN (not a microwave oven) at 375 for 90 minutes. Allow the roots to cool. Wrap each root in aluminum foil and place in your freezer for future use.

To fully appreciate what you have waiting for you and your family... take one sweet potato while still warm, make a cut in the form of a cross and ever so gently press the ends toward the center. Add a pat of cow's butter (do not remove the peel). Now, sit down, cross your legs and enjoy a delightful repast.

Sweet potatoes are not fattening, plus a medium-sized sweet potato provides more than the recommended daily allowance of vitamin A.

Any sweet potato recipe the American housewife wishes to prepare may—and should—be started with BAKED sweet potatoes. Dr. George Washington Carver recommended baking sweet potatoes over all other methods of preparing this vegetable almost 100 years ago. Unfortunately, this advice has not reached many cooks, yet. The only time you should use your microwave oven for sweet potatoes is to warm them up after they come out of your freezer.

I strongly recommend that you place your order early (today!). We want to ship early so we can get ahead of the cold weather in the northern parts of the country. Please make your check payable to Whatley Farms Inc.

For some truly down-home sweet potato recipes, you should contact:

Mrs. Faye Tilley
Tilley Farms
Rt. 16, Box 575
Cullman, Ala. 35055

When you request your sweet potato recipes from Mrs. Tilley, you should enclose a check for $2 and a stamped, self-addressed envelope.

You should always keep in mind that there is NO ideal place to store fresh sweet potatoes in the American home. The ideal conditions for storage are 55 degrees and 70 percent relative humidity. Under no circumstances should you store your sweet potatoes in your refrigerator. If you follow these recommendations, you will not lose a single sweet potato.

I would appreciate a note from each of you, telling me the price of sweet potatoes per pound in your local supermarket and how they compare with Whatley Farms'.

May the Good Lord Bless and Smile upon you.

Sincerely,

Booker T. Whatley, CEO
Whatley Farms Inc.

Whatley Farms did very well with that sweet potato offer when it was mailed to subscribers of our *Small Farm Technical Newsletter.* We sold more than 1,000 boxes of sweet potatoes. Another time, I did the same thing with pecans, the 1,200 pounds of nuts from the three big pecan trees in my backyard. I offered them at $2 a pound, with a minimum order of 10 pounds. Sold every single one. I even got an order from a guy in California who grows nuts. He grows English walnuts, pistachio nuts and macadamia nuts, but he didn't have any pecans. He bought 10 pounds. It's amazing what you can sell!

These customers are not looking for "bargains." They are looking for unique, high-quality and contamination-free foodstuffs. And they're willing to pay for them and to pay well. So give them only the very best products and service possible, or you're going to have all of your customers mad at you, as well as possibly the federal government. The

Federal Trade Commission has a thing called the 30-day rule. What that means to you as a direct marketer is that you have 30 days to ship the items ordered. If you can't do that, never mind the reason, federal law says you must:

• Notify your customers in writing that there will be a delay.
• Advise them of the approximate delivery date.
• Inform customers that, if they don't want to wait any longer for delivery, they have the right to cancel their orders and receive full refunds.

Leave too many customers holding the bag too often and the FTC can get downright nasty, slapping scofflaws with civil fines of up to $10,000 for each violation of the 30-day rule.

You shouldn't have to worry about any of that, though. Just don't promise anything you can't deliver on time and in perfect condition. After all, treating your customers right is in your very best interest. The better they like you and your goods and services, the more they will buy from you. There's a fantastic mail order market out there for many of your farm products. All you have to do is cultivate it carefully and watch it grow.

Be A 'Price Maker,' Not A 'Price Taker'

BATH, Pa.—For three generations, the Seiple family farm raised southeastern Pennsylvania's standard fare of potatoes, corn and wheat. That all changed in 1980 when Dave Seiple, then 30, and his brother, Daniel, decided that crop diversification and more aggressive marketing could offer the farm greater stability and profit.

Today, the 750-acre operation still produces the same field crops as its neighbors. But it also boasts 40 acres of U-pick strawberries, raspberries, peas, string and lima beans, cole crops, pumpkins and sweet and ornamental corn. As many as 800 customers a day, including occasional wholesalers, flock to the farm to pay the Seiples' prices, which are comparable to those at the supermarket. "Be a price maker, not a price taker," Seiple quips. "The market is there. It's just a matter of developing it."

Seiple should know. The experimental 2-acre strawberry patch he planted in 1980 has broadened to cover 15 different fruits and vegetables. "Strawberries are a great crop to start with for intensifying farming on a few acres," he says. "We had such good luck the first year . . . that we had to stop picking every other day so there'd be enough berries to meet the demand."

"Try one of my sweet raspberries!" There's always something ripening in the Seiple farm's selection of 15 different U-pick fruits and vegetables.

Seiple takes time out to give a strawberry picker some personal attention.

Seiple's early-season strawberries ripen just as most of the farm's field crops are seeded. "It's a nice time for cash flow," he says. "With everything in, it's good to get some money in. Our goal is to work up to 12 acres of them."

He's also added three types of peas, which ripen at the same time as strawberries, a feature he takes time to point out to berry-pickers. Seiple uses the same "while-you're-here" strategy a few weeks later, when string beans are ready. "Hey, raspberries are ripe now, too," he'll shout.

Profit Begins At Home

Seiple's experiment in diversification has blossomed into an even more important goal: to pare down the amount of rented acreage and focus money-making on the 210 home acres. Some unexpected progress was made in that area when a 10-acre housing development was built on part of a 20-acre rented tract. More farmland lost to development? Maybe, but Seiple prefers to regard the new neighborhood as a source of more customers. "That's my idea of farm preservation," he says.

"We think we're getting further ahead now," Seiple adds. "We're getting more money in, and we've learned a lot more about what we're doing with these (new) crops. Corn and soybeans are easy to grow. It's 'Bang,' put them in and hope you get a good price.

"U-pick vegetables are a better way to go. In the past two years, they've earned 15 to 20 percent more (than before). They take more labor, but the market's expandable. If rent for our (540 acres of) corn and soybean ground goes much higher, we'd rather pull back and intensify on our own land."

That's quite a contrast to the talk he hears as a board member of his county soil conservation district, two farm marketing cooperatives and the marketing committee of the Pennsylvania Farmers' Association. "I talk with a lot of farmers. At just about every meeting I go to, I hear horror stories about farmers cutting each other's throats, underselling each other, the co-ops. They don't stick together."

The best way to skirt such problems, says Seiple, is "to gain control of our products as close to the consumer as we possible can."

Meeting The Market

For Seiple, the first step toward that goal was the 1980 opening of a professionally designed, $100,000 retail store on the two-lane highway in front of his farm, 4 miles from the twin cities of Allentown and Bethlehem (combined population 175,000). "We knew we had pretty good traffic out there," he says, waving his clipboard toward the road.

The market, open year-round, was built to sell potatoes and some of the eggs from Seiple's 16,000-hen operation directly to retail customers. "It was a start," recalls Seiple. "We still hope to eventually sell all our eggs and even buy more if we have to. They're cheap."

To attract customers, the store also offered deli and bakery specials, canned goods, submarine sandwiches and frozen items. As the farm diversified, other farm produce made its way onto the shelves.

U-pick peas proved particularly appealing after Seiple added free shelling, performed by two, electrically powered cylinders costing about $1,200 apiece. The shellers dump empty pea pods through the barn floor to feed his 25 hogs housed below. The "free" feed from various vegetable wastes helps make the animals a profitable addition. Some hogs are sold as sausage or whole at the market, which also carries beef halves, quarters and hamburger produced from the half-dozen steers Seiple finishes each year.

"We didn't know whether to develop the store first, and then our produce, or the other way around," he says. "But we started from scratch. We're ending up improving both the retail and U-pick businesses at the same time. For the first time this year, we're giving U-pick customers discount coupons for items at the store. I guess we should have done this before."

The store employs five people full-time, and still caters primarily to local residents.

The pea shellers are among the few pieces of equipment Seiple had to buy to diversify the farm. "I already had the tractors, the cultivators, the sprayers for the potato fields," he says. "Everything's set up for two or four 36-inch rows, so I used what I had to fit my rows to my equipment."

He did, however, have to buy a two-row transplanter for $800. Two heavy-duty, electronic scales cost $715 each, but Seiple figures they quickly paid for themselves by measuring in increments of one-hundredths, rather than quarters, of a pound. The two electronic cash registers, bought for $631 each, keep good sales records and figure out change for busy field workers, a real help in the peak season when pickers are lined up. He bought all the equipment used, usually at auctions advertised in local newspapers. The Seiples built the outdoor weigh station themselves.

"It's easy to expand something like this without spending a whole lot of money," Seiple says. After five years, the May-through-October U-pick operation can attract 800 people on a good day. A fair-weather Sunday in October attracted 3,000 pumpkin pickers.

Finding and keeping reliable labor remains the most frustrating obstacle. "If kids do work, it's when it's convenient for them," Seiple says. "I get some older folks, too, though." But apart from the store and a few full-time field workers, Seiple has been able to get by with short-term, seasonal help.

25,000 Smiles

As the U-pick operation picked up steam, Seiple adapted other marketing features he'd read about and seen on other farms. "Everything you see here is a copy," he says, pointing out the proliferation of signs, the placement of the field's weigh station near ripening crops, and the strategic layout of the parking areas right in the center of the U-pick fields. "But I think I had one good original idea: the haunted barn at pumpkin time."

Seiple seized the opportunity to squeeze extra income from the farm's oldest, creakiest barn by converting it into a winding, hay bale-lined funhouse designed to attract more families during pumpkin season. "I had to offer them something they (other U-pick pumpkin operations) didn't have."

On weekdays, an odd assortment of hay-stuffed characters and "pumpkin people" dressed in skirts and overalls beckon passers-by to stop in. But on weekends . . .

. . . . On weekends, booing ghosts fly out in sheets to greet customers and direct traffic in the parking area, as noisy vampires roam the grounds to the delight of visitors.

"Haunted barn" has attracted up to 25,000 customers to the Seiple farm in the fall.

But the real attraction is inside the haunted barn. Seiple pays a local church youth group $100 a weekend to animate ghoulish Halloween scenes. While parents roam the fields for the family jack-o-lantern, youngsters pay 10 cents apiece for a tour of the old barn. Deep inside, witches are stirring bubbling cauldrons. An — ah — incompetent crew of doctors and nurses, dressed in full surgical attire, hover around one unlucky patient of whom they've made an awful mess. Coffins missing their lids are not so still, and hanged bodies swing from their ropes.

For the very young, impressionable set, Seiple has collected ducks, geese, rabbits, ponies and chicks that he borrows — or rents — for the pumpkin season.

"Last year, the weather stayed nice all through October," he says. "We figure we ran 25,000 people through here for pumpkins." And, yes, you can bet the barn will be haunted again this year and for many years to come.

Seiple uses every tack he can to lure customers into making the 4-mile trek to the country. Billboards rented throughout the area advertise his U-pick "hotline," which updates prospective pickers on prices and crops ready for harvest. "We want people to call, because conditions change from day to day," he explains. "When we're not open, we connect the hotline to a recording machine. But when we're here, I want people to talk with people. In the peak season, both phones are ringing all the time. We definitely need it. I think (the hotline) is paying its way."

He also places newspaper ads and airs radio spots. "It's hard to gauge where the advertising pays," he says. "I think sometimes we too often take the shotgun rather than the BB gun approach." Local newspaper photographers have learned there's always a good seasonal picture to be taken at his place. The Sunday papers often feature photos of local children biting into a big, juicy Seiple strawberry, or lugging away a Seiple pumpkin that's almost as big as they are. "A little publicity like that always helps," Seiple says with a grin.

Because they are removed from farming by several generations, most of Seiple's customers "like the idea of coming to a real, live working farm. We may move from selling the product to selling the experience. We've tried some of this lately in our advertising, and it seems we're getting a lot more families in here, a lot more kids," Seiple says.

New Crops, New Profits

Seiple has all but gotten out of field work these days. "You can always find someone who can drive a tractor," he notes. He concentrates on detailing work for the help, looking after customers and making plans.

His work with the co-ops and the state marketing committee helps him keep track of market trends, competition and prices. "Pretty soon, U-pick strawberries are going to be saturated around here," he predicts. "By then, I can expand on my raspberries and blueberries. It'll take a couple of years before all these other strawberry growers back off." But for now, Seiple can afford to hold out for his price on most crops. "I just got a call from the produce man at one of the local supermarkets offering me a dollar a quart for strawberries (Seiple's U-pick berries sell for $1.25 a quart). I turned him down."

Seiple says he'll never focus on unprofitable crops like peppers and tomatoes. "Anybody can grow them in the backyard. I raise a few just for people whose peppers got eaten by the rabbits." Speciality crops are not in his plans, either. "People around here won't go for them," Seiple says matter-of-factly. "It helps to look at your ethnic groups. Grow what they want. People will give you suggestions."

The key to choosing a new crop is its labor requirements, he notes. "That's why I try to keep good records on the time devoted to each crop. This is a labor-intensive operation." The one crop he's taking a hard look at adding now is table grapes. "They're different," he says, pointing out that the new seedless varieties lend themselves well to both U-pick and retail sales.

Greenhouse production is out. "Other retailers around have them now, and if I did, too, we'd run into the same problem of overproduction. Besides, I don't have the (horticultural) expertise," he says.

Overall, Seiple is confident that his plans are paying off. "We borrowed a lot of money, and we're now digesting what we bit off," he concedes. "We spent money where we had to, and the cash flow's getting better.

"Right now, we're still looking to the future," he adds. "But the opportunities are much greater."

He pauses to glance at one of the wooden barns crying for paint. "Yeah, I'd say we're a lot better off than we were before. I think we've gone the right way. Being one of the first is the key. Those who come in after me will now have to meet my standards."

Editor's Note: *This article first appeared in the Sept./Oct. '85 issue of* The New Farm, *shortly before Dave Seiple's sudden death. Although Dave is no longer with us, his story belongs in this book because, when it came to diversifying crops and marketing directly to the public, Dave Seiple had an uncommon amount of common sense. The real tribute to his foresight and skill, though, is the fact that the Seiple family is carrying on the innovative production and marketing practices that Dave began. We wish them all the best.*

Let Your Software Do The Selling
So you can take care of business *and* the farm

RON CASTLE

If farmers in middle Tennessee are typical of farmers in other parts of the country, my guess is that most continue to be skeptical about the arrival of the computer age on the farm. Like sizing up a good bull or a new piece of machinery, the big question is: "What's it really going to do for me?"

In a word, *plenty!*

With the help of a computer, our direct marketing efforts in '84 resulted in the sale of 16 lambs, for a total of $1,750, and three 850- to 900-pound steers for about $1,850. If we figured time spent on our marketing at $10 per hour, this expense would amount to less than 3 percent of our net sales. Our customers are getting a good deal, and we are getting a guaranteed $1 per pound for our lambs, and 75

MARKETING, THE MOST IMPORTANT PART

Ron Castle at the keyboard of his Kaypro 10 personal computer, which he used to write this article.

cents per pound for our steers; much more than the average wholesale market prices in our area.

We have developed a friendly, informed clientele over the course of three years, and personal attention, made more feasible by the computer, is partly responsible. Now, most of our new business comes from customer referrals, and we are growing steadily to meet the demand.

We use our personal computer to communicate with our 50 or so regular customers who buy freezer beef and freezer lamb directly from us. Most of our customers live in Nashville, about 70 miles away; Chattanooga; or outside our immediate area. Calling long distance would be costly and time-consuming. Being part-time farmers, we don't have time to call when we should, or write letters by hand.

By using a simple data base program (a computerized filing system) and a word-processing program, we keep up a regular and personalized correspondence with our customers that we would never be able to do otherwise.

We also use the data base to record how each customer likes his/her beef or lamb processed. This can get complicated at times, since we sell lamb either by the half or whole and beef by the quarter, half or whole. We use a custom slaughterhouse, James' Meat Co., to process our animals, and they accept our computerized processing instructions, rather than their standard forms. That saves both us and them a lot of time, since we don't have to write up extra processing orders, and they don't have to fill out their usual receipts. We also use the computer to match customers whose processing requirements are about the same.

Here's an example of how we use our computer for marketing. In May, we had a lamb barbeque to promote the idea of eating and buying lamb. This was our first lamb crop, and we had never sold lamb before. Using the data base, we printed out mailing labels for all of our customers. That took about 10 minutes. Then, using the word processor, we wrote and printed about 50 single-page letters inviting our customers to the barbeque, telling them about the lamb we would have for sale, and encouraging their advance orders. We could have printed only one letter and run copies, but the computer-driven printer saved both a trip to the copying machine in town and $5 for the copies. Plus, the letters were neatly printed and each one personally signed. Printing out the letters took about an hour, and getting them in envelopes took maybe 15 minutes. With a walk to the mailbox, the entire job was done in less than an hour and a half.

A few days after the barbeque, we sent thank-you letters and ordering information to the folks who came, and a different letter to those who didn't come, telling them what a great time we had stuffing ourselves, and how they could order *their* lamb. Each letter also had a P.S. at the bottom, advising that we would have three steers finished and ready for slaughter in early September and that reservations for beef were also recommended. This job also took about an hour and a half.

As the orders came in, we recorded them on a customer list printed out from the data base. When our first group of lambs went for processing, we used the word processor to prepare the instructions for the butcher, and again to write a standardized letter to our customers to let them know when they could pick up their lamb. It took about an hour to get all of this done and in the mail. The processing instructions were delivered with the lambs to the slaughterhouse.

Our use of the computer isn't limited to sales correspondence. We are also using the data base program to keep our flock registry on our Suffolk sheep and production records on our breeding ewes. Eventually we would like to keep our farm accounting on computer, but so far, manual accounting hasn't been too time-consuming.

Buying Software and Hardware

Many books and magazines are available on buying personal computers, so I will leave you with only two recommendations: Find the software you need first, then find the hardware. After a year of studying and shopping, we finally decided to buy a Kaypro 10 computer (Kaypro Corp., 533 Stevens Ave., Solana Beach, Calif. 92075), because it came complete with word-processing, data base and other programs. A dot-matrix printer and a few minor accessories were needed to make the package complete. We decided we needed a better data base program than the one that came with the Kaypro, so we purchased the "Friday!" program (by Ashton-Tate Corp., 9929 W. Jefferson Blvd., Culver City, Calif. 90230). We use Friday! to keep track of our customers, their processing requirements, and the records on our sheep. Our total investment amounted to about $3,500, because we bought the top-of-the-line Kaypro.

That was back in 1984. (We bought a second Kaypro in '85.) Computers have improved tremendously since then. Prices have come way down, while computing power, memory capacity, and software selection and quality have all increased exponentially. Today, you could get started nicely for less than $2,000 for a computer like an Apple IIe with a near letter-quality dot matrix printer.

Computer Tax-Deductible

If you are buying a computer to use in your farming business, it can be treated as a depreciable asset for tax purposes. Depreciation reduces the amount of your gross income that is subject to income tax.

But, if the computer is also used for non-farm activities like playing video games, preparing the children's homework and maintaining Christmas card lists and recipe files, tax laws require that you keep a log of the amount of time your machine is used for business.

The answer to "What's a computer going to do for me?" has been "plenty." While I could probably market direct without the computer, I'd prefer not to try.

They Need Chemical-Free Food
Some people just like organic produce, but a special group of consumers literally can't live without it.

CLINTON, Mich.—A neighbor was the first to buy Jean Winter's organically grown produce and grains for the simple reason that the chemicals in most foods literally make him sick. He told some friends with the same problem about Winter's 96-acre farm. They told some friends and soon, a lot more customers started coming to Winter's Eden Acres farm here.

"As we get more acquainted with this group, I think our business will build up enough so that we'll go into farming full time," says Winter.

The "group" she refers to is the Human Ecology Action League (HEAL), a non-profit group formed in 1976 to help the thousands of chemically sensitive people throughout the United States and Canada. Winter is working with Ann Arbor HEAL members and an area doctor who treats 500 to 600 people with chemical sensitivity problems.

"Pesticides are the number one most hazardous exposure to this group of people," says Dr. Theron G. Randolph, a Chicago physician who pioneered the study and treatment of chemical sensitivity more than 40 years ago. "The agrichemicals are disastrous..."

Major agricultural areas like California's San Joaquin Valley and South Florida are almost out of bounds for the chemically sensitive, because of the large amounts of pesticides sprayed on farm fields, golf courses and homes much of the year, he says.

"There is an incredible use of pesticides here. I don't know how you can avoid it," says Dr. Hobart T. Feldman, an allergist with offices in Hollywood, Fla., and Miami. He treats scores of people with severe reactions to pesticides in food, water and air. "You just hope you don't get a reaction and can continue living here."

But the problems of widespread pesticide contamination and human sensitivity are hardly confined to California and Florida.

"I have patients from virtually every state and 16 foreign countries," says Dr. Kenneth Krischer of Fort Lauderdale, Fla. His clinic attracts chemical sufferers from France and Australia. "More and more people are becoming sensitive by being exposed to chemicals. It's fairly widespread."

That, coupled with the fact that many susceptible people can only survive by eating foods grown without pesticides and avoiding as many other chemicals in the environment as possible, creates a ready market for organic farmers. Raymond and Bernice Rus, for example, maintained their 7-acre organic fruit and vegetable farm in Hypoluxo, Fla., for some 30 years largely by selling directly to people with chemical sensitivities throughout Palm Beach and Broward counties. The demand for organically produced beef, then pork, lamb and poultry by HEAL members around Washington, D.C., had a lot to do with turning the Garnett Coordinated Biofarm System into a $250,000-a-year operation in just eight years, according to founder Gwynn Garnett.

Despite such a demand for chemical-free foods, there has never been any major effort to put the chemically sensitive in touch with organic farmers around the country. Only HEAL chapters in California and around Chicago have tried to bring producers and consumers together.

"We've never gone looking for them," says organic farmer Judy Yaeger. Although Yaeger now has only a handful of such customers and remains somewhat skeptical of the chemical sensitivity problem, she's convinced of the customers' commitment to buying only organic foods. Some HEAL members drive 150 miles from their homes around Chicago to buy produce at Yaeger's Lawton, Mich., farm.

"There seems to be an upsurge (in such trade)," Yaeger adds. "I don't know if more people are just finding us or if there is more of a problem with (chemical sensitivity)."

Winter believes both factors are helping build the market. Many people with chemical sensitivities have to shun not only pesticides, but the growing number of chemicals in common household cleaners, new furniture, synthetic fabrics, plastics and paper products. "If you stop and think about it, it's mind boggling the way they have to live. Word of mouth is how they find out about organic farmers and it spreads like wildfire, they're so desperate to get food. These people will find organic farmers — if they know where they are.

"Now they're trying to establish a nationwide network... a computer network of organic farmers, growers, fiber sources and medical help," she adds.

Doctors who work with the chemically sensitive believe this could be a real benefit to their patients, many of whom are sickened by just a few molecules of pesticides and other chemicals in the environment.

"Symptoms begin rather quickly, almost with the first breath or two as they walk into a room. Some people develop foggy brain, cerebral edema, go blank... their mind doesn't function as well as it should," says Dr. Joseph B. Miller of Mobile, Ala., a member of the Food Allergy Committee of the American College of Allergists.

Eating, drinking or breathing minute traces of many chemicals cause drastic, sometimes bizarre physical and emotional changes in susceptible adults and children. "Many people are this way chronically," Miller says. Yet they have no idea what is making them sick, unhappy or nervous.

Research on the effects of pesticides on humans is in its infancy. No one can say for sure how or why chemicals affect some people and not others. Some doctors blame stress. Another theory is that pesticides disrupt the operation of the hypothalamus, a gland with important influences on the nervous system.

Randolph blames the more than 50,000 pesticides and synthetic chemicals forced on mankind since World War II. "The human race has had millions of years to adjust to cold, light and other solar radiations. Most of us are adequately adapted to common foods, plant pollens, and oils, spores, insects and other animals. In contrast to these naturally occurring physical and biological exposures, the man-made environment is a relatively new exposure."

For years, Randolph's work was largely ignored by the medical establishment. But the many pesticides and other harsh chemicals routinely used in hospitals made chemical sensitivity among doctors, nurses and housekeeping personnel almost as common as it is among chemical manufacturing plant workers, commercial pesticide applicators, farmers and farm workers and others continually exposed to pesticides, according to some government and private research. In desperation, other doctors began coming to Randolph for treatment. The chemical sensitivity problem in doctors was so serious that in 1967, a group of doctors from around the country founded the Society for Clinical Ecology. It now has more than 350 members.

The plight of the chemically sensitive has even been officially recognized by the Internal Revenue Service. Since late 1976, the added cost of maintaining a chemically free diet under a doctor's orders has been a legitimate tax deduction.

For more information on chemical sensitivity, write: HEAL, 2421 W. Pratt, Suite 1112, Chicago, Ill. 60645.

Chemical Sensitivity Her Personal Nightmare

BOCA RATON, Fla.—All 10 doctors had given up on Janet Mikkelson.

In five years, not one had come close to finding the cause of her deepening depression, mental brownouts, general lethargy, breathing troubles and nausea. In desperation, they referred Mikkelson to a psychiatrist. Maybe her problems were psychosomatic, the doctors theorized.

But one year later, the psychiatrist was as baffled as the doctors. And Janet Mikkelson was no better.

Then the psychiatrist attended a lecture at the University of Miami, and the missing pieces of Janet Mikkelson's bizarre medical jigsaw puzzle suddenly fell into place.

The lecturer was Dr. Douglas H. Sandberg, a University of Miami School of Medicine pediatrics professor. Sandberg's specialty is digestive and nervous disorders, but the lecture was on his hobby—the little understood, often ignored effects of extreme chemical sensitivity.

The psychiatrist recognized all of Mikkelson's symptoms and sent her straight to Dr. Sandberg.

"He saved my life," Mikkelson says of Sandberg.

The affliction that had stumped the other doctors was nothing new to Sandberg. He treats hundreds of adults and children with the same problem. They come from throughout South Florida.

Chemical sensitivity, Sandberg explains, is not a simple allergy as many sufferers and doctors mistakenly believe. It is the result of long-term chemical poisoning—continued exposure to increasing amounts of poisonous chemicals in food, air, water and the environment.

"Pesticides are high among the chemicals you're exposed to, particularly in South Florida," he says. "The susceptibility of some individuals is so high that even minute amounts (of a pesticide) make them seriously ill. Some people are not able to tolerate their house indefinitely (after it has been fumigated for termites)."

Just as some people physically cannot stand alcohol or cigarette smoke, Sandberg says others have an inherent sensitivity to chemicals. And, once they are made sick by one chemical, they usually are affected by others.

"Usually, people who are super sensitive to foods are super sensitive to chemicals . . . a broad range of chemicals," the doctor says. "Many everyday items—chlorine bleach, disinfectants, bug sprays, household cleaners and polishes, even diesel exhaust—cause serious problems for them."

Mikkelson is living proof of that. "If I have an exposure to a chemical, whether I eat it or drink it or breathe it, I get sick," she says. "If I walk into a grocery store and they have just sprayed it (with pesticides), I get totally wiped out."

It's not uncommon for Mikkelson to "blank out" in a freshly sprayed store. She becomes disoriented and finally wanders outside, staggering as if she were drunk.

"One bite of anything containing chemicals and she is right around the bend," says David Hardwick, manager of the Harmony & Lotus natural foods store in Pompano Beach, where Mikkelson buys most of her food.

Many people like her shop at Harmony & Lotus, but Mikkelson's condition is worse than most. Store employees fondly call her "the Organic Lady." In fact, Hardwick sometimes has Mikkelson sample "organic" produce from new suppliers to really test its purity before setting it out for sale.

Because of her sensitivity to chemicals, Mikkelson quit her six-day-a-week bookkeeping job in a hardware store where she was regularly exposed to pesticides. She now does accounting at home, but has to wear a surgical mask because of the chemicals in the ink.

She quit smoking, gave up coffee, milk and beef, which can contain chemicals. Mikkelson drinks only bottled spring water, but has found that it—and all her foods—must be in containers made of glass, not plastic. Instead of killing palmetto bugs with Raid, she now swats them with rolled up newspaper.

"I am not alone," she adds. "I know many people who have symptoms similar to the ones I have."

Chapter 8
Beat The Weeds

Your Mower—The Best Herbicide Yet
Introducing a broad-spectrum weed killer that's cheap and effective.

Efficiency is the only way to make it in farming these days. For Virgil Bareham, who grows 150 acres of tart and sweet cherries near Sutton's Bay, Mich., that means getting two jobs done at once with his mower.

"I'm mowing anyway, why not get a little extra weed control and mulch in the bargain?" says Bareham, who has modified his mower to blow all of his orchard driveway cuttings into the tree rows. "Mulch has a lot of values. It holds moisture, smothers weeds and puts nutrients back into the soil."

Field-crop farmers can also turn their mowers into broad-spectrum herbicides, thanks to research into the life cycles of weeds, which has revealed when they're most vulnerable to cutting. Many weeds are best cut when they are in flower, but the timing of subsequent mowings can make the difference between eliminating the weed or allowing it to survive.

For Bareham, mowing and mulching has saved one or two $10-per-acre herbicide applications per season. It has also relieved water stress on his trees at critical blossom and fruit set times, and helped his trickle irrigation system provide moisture more evenly to his trees. Bareham grows the traditional, long-lived perennial ryegrass and Kentucky bluegrass for his driveway cover crops. He makes his first cutting in midspring when the grass is about 18 inches high.

"That machine lays down an inch of mulch just as fast as you can cut," says Bareham. His Woods 120 mower is a side-by-side tandem unit which he modified to give strong side-delivery of cuttings. First, Bareham changed the drive mechanism to reverse the direction of his left side cutter so that it feeds clippings to the right side blades. Then he cut away a portion of the mower body and installed a grass chute to direct the flow.

Don't reach for the cutting torch right away, though. Some mower manufacturers, including Woods, offer models with built-in or bolt-on chutes for safely directing cuttings where they're needed.

Mow/mulching smothers weeds, holds moisture and puts nutrients back into the soil.

Boosts Tree Growth

Bob Gregory, who with his brother, Don, manages the family's 650-acre cherry and apple orchard near Traverse City, Mich., says mow/mulching not only helps control weeds, but is crucial to the survival of costly new plantings. "We know that mulch increases the survival and growth rate of young trees. Blowing that mulch in is fast and easy. Where we don't have enough mulch, we'll apply hay by hand. It's that important." Gregory also uses a modified mower to distribute cuttings in an 8-foot-wide strip in the tree rows. Unlike Bareham, the Gregorys first mow orchard middles when the cover is 8 to 12 inches tall. "You want to cut it before it sets seed, so you're not blowing seed into the rows."

The Gregorys are still trying to find the best cover crop for the system. They have experimented with sorghum, oats, and *annual* rye. The rye is managed much as a traditional driveway cover, but oats and sorghum require some new and different approaches, says Bob. "With oats, you're seeding in late summer and letting the crop winter-kill. It leaves a mat that provides fairly good control and a nice layer of mulch until you plant again next year." With weed-fighting covers, Gregory has eliminated three of his usual four herbicide applications at a savings of about $20 per acre. He has increased his planting costs by only half that amount for a net saving of $10 per mow/mulched acre. But just as important to Gregory are those tough-to-put-a-price-on advantages: the survival and fast growth of young trees under mulch, recycling of nutrients to the dripline area, and the gradual improvement in soil tilth.

Natural Weed Killers

Researchers are finding that some types of mulches can pack more weed-controlling punch than others. Plants such as annual rye, sunflowers, sorghum, and oats release substances that are toxic to other plants, says Alan Putnam, a Michigan State University plant scientist. These toxins inhibit the germination and growth of weeds, benefiting the source plants and if the grower plays his cards right, the cash crop. "Right now, annual or grain-type rye looks like the best crop to use," says Putnam. "We have the most experience with orchard situations, but annual rye is doing a nice job in vine crops and certain vegetable crops, too."

Finding cover crops that could be grown for weed control, water conservation *and* nitrogen could tip the cost balance even further in favor of mow/mulching, Putnam says. "White clover does nicely as a nitrogen-fixing cover crop, but we haven't worked with it enough yet," says Putnam. One disadvantage of using a legume, he says, is that bees, which should be pollinating fruit trees, are distracted by tempting clover flowers.

Mow Pasture Weeds, Too

Mowing is one part of a total weed control program for pastures, field edges, and other parts of the farm where row crops aren't grown, says M.K. McCarty, a weed scientist with the University of Nebraska who has conducted a 20-year-long comparative weed control study. "One of the most effective ways to use the mower is to prevent seed production in annual weeds," he says. Annuals should be mowed when flowers *first appear,* because some weed seeds will manage to germinate even though the plant is cut soon after pollination.

McCarty says mowing is less effective on perennial weeds and low-growing weeds, which can set seed below the blades. He recommends wick application of systemic herbicides for persistent perennial weed problems.

But perennials can also be controlled by well-timed mowings, which can both prevent seed production and starve underground parts, recent research shows. To control tall perennial weeds, repeated and frequent cutting may be required for one to three years. Never let weeds replenish their stored food supplies. The best time to start mowing or cultivation is when the root reserves are at a low ebb. For many species, this is between full leaf development and the time when flowers appear during late spring. New stems can only grow by using up stored food.

In pastures, adequate soil fertility, drainage, and near-neutral soil pH can help desirable legume and grass species out-compete weeds. Controlled, rotational grazing, rather than prolonged set-stocking, can also suppress weeds and favor good pasture regrowth.

Providing bark-chewing rodents with cover is one possible disadvantage of mow/mulching. Most of the Gregory's trees go through winter, the riskiest season for rodent damage, with a layer of mulch. But routine anti-rodent measures are enough to prevent tree damage, says Bob. "We protect our younger trees with the white plastic spiral tree guards. In mature orchards, we don't leave crates or junk around, because that's the main rodent habitat. If we have to, we'll use poisoned baits. The trees are too big of an investment to leave the rodent problem to chance under any conditions. Maybe we're a little more conscientious now."

Can mulches tie up nutrients that should be going into the trees? "I go by the theory that grass mulches tie up a little extra nitrogen at first, but then gradually release it back," says Gregory. He applies about 2 pounds of actual nitrogen per tree in split spring-fall applications. As mulches decompose, they also return micronutrients to the tree root zone, he says.

Both Gregory and Bareham say they minimize the dangers of the mow/mulch system by not mowing in the vicinity of orchard workers. Reducing the risk of thrown objects is another reason to keep a clean orchard, they add.

It's also important to protect mower operators, who can fall under cutters. "In a blade contact situation, the tissue usually can't be restored. It often results in an amputation," says Wesley Buchele, Iowa State University agricultural engineer and an authority on mower safety. How to avoid such an accident? He rates a full cab as the best protection. Next is the use of tractor seat belt and roll cage.

Nearly every farm has a mower, and used wisely, it can do two jobs at once, like mowing/mulching. With good timing, a mower can be used to kill annual weeds before they can set seed, or starve out perennial weeds. "Herbicides work fine on certain weeds, but I can't find one to do the whole job like the mower," says Bareham.

Nature's 'Herbicides'
Science is finding they can be stronger than 2,4-D.

When plant pathologist Alan Putnam left his Michigan State University lab to check his experimental plots, he was surprised. Even though the plots hadn't been cultivated all summer and no herbicide was applied, not one weed emerged among the vigorous snap beans.

Near the cool, eastern shore of Lake Michigan, 150 miles to the north, Bob Gregory, who manages a 650-acre cherry and apple orchard, found only a few weeds when he checked the tree rows, even though he eliminated all but one of his usual four herbicide applications.

Putnam and Gregory are among the handful of researchers and farmers who are pioneering new uses for *allelopathy,* a biochemical phenomenon that can be more than twice as powerful as 2,4-D.

While Gregory still considers his 70 acres under allelopathic weed control to be experimental, he estimates that he is cutting his weed control costs by at least $20 on each of those acres.

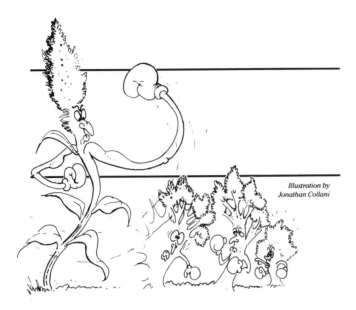

Illustration by Jonathan Collani

"We've gone from one residual herbicide application and three (herbicide) burnbacks to one burnback," says Gregory, who along with his brother, Don, manages their family-owned orchard in a cherry-growing region near Traverse City. "We've had our best success growing fescue and rye on the entire orchard floor and killing it back once under the trees. The rye mulch gives off its own 'herbicides' all season long," Bob says. "We're going with allelopathic weed control in all our new plantings."

But what exactly is allelopathy? Certain plants, such as rye, sunflowers, sorghum, oats and even some microscopic plants, release substances that are toxic to other plants. These toxins inhibit the germination and growth of other plants, benefiting the plants that produced them. Also, some substances released from plant leaves, roots and decomposing residues encourage special microorganisms in the soil which inhibit higher plants. The term allelopathy is derived from two Greek words meaning "mutual harm."

At Michigan State, Putnam and others are testing the allelopathic qualities of ryegrass and sorghum in rotation with vegetables. In the case of the snap bean plots, fields of sorghum were planted in the fall and allowed to winter-kill. The beans were then direct-seeded into the decomposing crop residues. This method also works with other vegetables, such as transplanted tomatoes and cabbage. Since rye is winter hardy, an herbicide was used to kill it before vegetable planting. Rye significantly reduced early season common ragweed and common lambsquarters, compared with control plots with no rye.

"If someone wanted to try it without using herbicides to kill back the mulch, they could try planting oats, sorghum or a sorghum-sudangrass hybrid in early fall, because these would winter-kill. The weed suppression would probably not be as good as with rye, however," says Putnam. Even though sorghum-sudangrass yields a lot of biomass for weed control, it has to be planted early in the fall in most parts of the country and, depending on the crop rotation, may not be feasible for some farmers, he says.

Putnam cautions against planting small-seeded vegetables, such as carrots and spinach, in a rye mulch. The allelopathic chemicals from the mulch will inhibit germination and growth of these vegetables just as they would stunt weeds.

Some Residues Risky

Researchers think the allelopathic action of some crop residues is due to a combination of toxins *from* the residues and from microorganisms caused to grow profusely by substances *in* the residues.

Allelopathy is a consideration in field crops, as well as orchards and vegetables. A 23-year experiment at the University of Nebraska-Lincoln shows that crop residues in a corn-oats-wheat rotation can produce powerful "herbicides." Some plots were subsurface-tilled (stubble-mulched) and some plowed. Decreased yields and abnormal crop appearance occurred in the stubble-mulched plots in years with normal to above-normal precipitation.

Solutions made with fungi from soil in the stubble-mulched plots reduced corn seed germination. One fungus, *Pencillium urtieae,* produced a toxin called patulin, which is particularly toxic to corn, wheat and other higher plants. Patulin reduced corn germination by 85 percent, while 2,4-D reduced germination by only 40 percent.

In other studies, extracts of roots and tops of alfalfa, red clover, birdsfoot trefoil, timothy, bromegrass, orchardgrass, and reed canarygrass were tested against seed germination and seedling growth of the same group of plants.

Overall, grass extracts had less effect than legume extracts, and extracts made from plant tops have greater effects than those made from roots. Grass extracts are less toxic to grass species.

Some extracts cause twisted roots, prevent root hair development, and kill roots. The following species are rated in order of decreasing herbicidal effects: alfalfa, birdsfoot trefoil, ladino clover, red clover, reed canarygrass, bromegrass, timothy, orchardgrass. Alfalfa's pronounced ability to inhibit all of the test species may explain alfalfa's dominance when grown with other forage crops. In fact, alfalfa even becomes toxic to itself, which may explain why alfalfa stands thin out after a few years and why it is so difficult to establish a new alfalfa seeding in an old alfalfa field.

This effect may also be responsible for what farmers over the years have come to call "soil sickness," an as yet unexplained condition that plagues continuous monocultures.

On-Farm Research

In orchards like Bob Gregory's, these discoveries are being put to work. Gregory drills or broadcasts rye into fescue. He drills two bushels per acre, or broadcasts up to three bushels per acre. Either way, he says, tillage is a problem.

"Even when I broadcast, I have to harrow a little to cover the seed. This can expose some roots, especially on older trees that have pushed big roots near the surface," Gregory explains. These exposed roots produce undesirable sucker growth and increase the chances of disease. "I wish somebody would invent a no-till drill that's only 18 inches high. I could drive it right under the trees," he says.

Besides the rye burnback method, Gregory grows sorghum in orchard driveways, then blows cuttings under the trees as an allelopathic mulch. "The sorghum was erratic this year," he says. "It was too dry to get enough growth, enough organic material to blow under the trees." He also dislikes tending a crop in the driveways. But the sorghum mulch method may someday replace synthetic herbicides in orchards if the problems can be sorted out.

Whether allelopathy will affect the trees, themselves, is not yet known. Gregory says he'll continue to experiment until he or the researchers find out.

More Weed Fighters

Here are more plants that are being studied for their allelopathic properties:

White lupine (*Lupinus albus* L.), a high-alkaloid winter cover and green manure crop which acts as a natural herbicide when the plant material decomposes in the soil.

Sunflower debris, when applied to greenhouse and field soils, stunts growth of redroot pigweed.

Oats, peas, and buckwheat suppress growth of lambsquarter. Buckwheat is very competitive for nitrogen, phosphorus, and potassium, thus suppressing weeds. Oats emit an as yet unidentified allelopathic compound.

Barnyardgrass and giant foxtail reduce corn yields, but giant foxtail and corn residues increase soybean yields.

Yellow nutsedge residue inhibits soybeans, barnyardgrass, and pigweed, but does not adversely affect corn.

Even weed *seeds* seriously inhibit crop growth, researchers have found. Autotoxicity, "self-poisoning" due to substances in seed coats, is partially responsible for some seed dormancy, also. Breeding programs for crops such as sorghum take this into consideration. Plant breeders use this property to develop seeds that store well after harvest. There is also some evidence that allelopathic inhibition of seed-rotting microorganisms allows a seed to resist decay as it lies in the soil. Further information on allelopathy may be found in the book, "Pest Control With Nature's Chemicals: Allelochemics and Pheromones in Gardening and Agriculture," by Elroy L. Rice (University of Oklahoma Press, 1983). The book costs $28.50. Rice's earlier book, "Allelopathy," (Academic Press, 1974), is out of print, but should be available at many libraries.

Kick Up Your Yields

New research proves that cultivation puts more bushels in the bin and more money in your pocket.

Just one cultivation each season is all it takes to increase yields as much as 24 bushels per acre in corn, and nine bushels in soybeans, even on ground that already has good weed control, a three-year University of Illinois study shows. Cultivation increased yields in nearly all the tillage/herbicide combinations tested.

"This was no fluke of the weather. We had one wet year, one dry year, and a normal year. Cultivation worked each season," says Marshal McGlamery, a U of I agronomist who conducted the study with John Siemens, an agricultural engineer.

More research revealing that cultivation is a yield booster comes from Ohio State University. "The effect of cultivation on yields is most noticeable on our soils that seal over after a rain," says Don Eckert, an OSU agronomist. "We do recommend at least one cultivation, even if there is fairly good weed control with herbicides. You can pick up seven to 10 bushels with one cultivation, no problem. It means money to you."

Louisiana State University studies also show that cultivation increases soybean yields on crust-prone soils by an average of 18 percent (3.7 bushels per acre). More importantly, cultivated soybeans netted $16 per acre more than non-cultivated beans.

Agronomists have estimated the per-acre cost of a row crop cultivation to be about $4, and a ridge cultivation, $6. "In contrast, many of the newer broadcast post-emergence herbicides are selling at $15 to $20 per acre for a single chemical," says Richard Johnson, an agronomist for Deere & Company. "More than one herbicide is needed if a variety of weeds is present." Add $2 per acre to that for application, and you have an accurate picture of herbicide costs.

Cuts Erosion, Too

"The fact that cultivation improves yields has been ignored by everybody but the farmer," says Johnson. "The feeling among many researchers has been that, if you can achieve good weed control with herbicides, there's no need to cultivate. But farmers haven't thrown away their cultivators." Johnson raised some academic eyebrows with an article on cultivation in the June/July '85 issue of *Crops and Soils* magazine, citing research from across the country which shows cultivation not only increases yields and cuts costs, but *decreases* soil erosion, too. For example, a Purdue University study stressed the importance of cultivating after hard spring rains to eliminate surface crust and improve water infiltration. The researchers found that a crust may have a greater effect on water intake than would soil type, slope, moisture content or soil profile.

Where soil cracks appear during dry periods, shallow cultivations are recommended to close the cracks, produce a dry surface mulch that reduces soil moisture evaporation, and produce an irregular soil surface, says Johnson. Cultivation can increase water infiltration and delay runoff, even in conservation tillage systems. If the site is especially erosion-prone, cultivator shovels can be adjusted to throw very little soil, allowing most residue to remain on the surface. New ridging and non-ridging cultivators are available to work in heavy residue.

Cultivation Prevails

The latest word on cultivation was presented to the 1985 summer meeting of the American Society of Agricultural Engineers by Siemens and McGlamery. The main purpose of their study, which was sponsored in part by the Shell Chemical Company, was to assess preplant incorporation, pre-emergence herbicides in relation to moldboard plowing, chisel plowing, disking, no-tillage, and row middle cultivation. Each year, one set of the 40- by 270-foot (16-row) plots received a single, early June cultivation *in addition* to the various herbicide treatments being investigated.

Excellent weed control is just one advantage of cultivation. It also cuts herbicide costs, boosts yields and improves water infiltration.

The U of I cultivations were done about one month to six weeks after planting, when the corn was about knee-high, says McGlamery. Most of cultivation's benefits can be obtained by cutting only deep enough to break up any soil crust and eliminate weeds, says Deere's Johnson. Running too deep and too close to the row during later growth stages can prune roots and reduce yields.

Cultivations produced some unexpected high yields. "These are poorly drained soils, but they don't have a tendency to seal over, which would favor cultivation," says Siemens. "In all but one of the 20 comparisons, cultivated plots yielded more than non-cultivated plots." Part of the yield increase may have come from improved weed control, but most was due to improved aeration and a higher water infiltration rate, say the researchers.

Why not cultivate two or three more times and eliminate the herbicide costs altogether? "There you're taking a chance on not reaching the weeds in the row," Siemens says. "The weather can throw your plans to cultivate, too. One compromise is to band herbicides into the row and cultivate, too."

Ridge tillage and cultivation, which *can* bury weeds in the row, were not part of the Illinois study. A standard shovel cultivator was used on the tilled plots, while a rolling cultivator was used on the no-till plots.

If Siemens were actively farming, would he cultivate, even if he were applying herbicides? "I'd surely try it," he says. "We're recommending that farmers cultivate some check strips."

Apparently, the benefits of cultivation are becoming better known to the financial sector, too. "I say no a lot more than I say yes," one frugal bank president told the *Des Moines Register* in a story about the large number of banks experiencing difficulty with farmer debt. He approved only one equipment loan that season: $1,000 for a used cultivator!

Chapter 9
Beat The Bugs And Other Pests

A 'Natural' Answer To Antibiotics
It may be just the thing to help control livestock diseases safely and cheaply.

Remember those robust, 100-year-old European peasants who appeared in TV commercials for a famous yogurt manufacturer a few years ago? While their long, healthy lives may not have been totally due to eating yogurt, scientists say it just could be possible that certain bacteria in cultured milk products help control disease naturally.

And when relatives of those bacteria are fed to stressed livestock, the result can be improved disease-resistance and feed digestion, and a nearly 50-percent reduction in health-care costs per head, by some estimates. "It's a totally different approach to animal production," says Roger Crum, vice president of marketing for Pioneer Hi-Bred International's Microbial Genetics division. "Our products are based on naturally occurring organisms. They're used to get (animals) back on feed . . . And help them provide their own defense system."

That's not to say that livestock farmers should throw out their Bovatec and Rumensin and stock up on La Yogurt, Crum quickly points out. "If an animal is sick, it's going to have to be treated with something." But if the health problems caused by today's intensive, confinement livestock operations can be largely *prevented,* Crum and others feel that the need for antibiotics and drugs will be reduced. And that's exactly what probiotics, a new generation of microbial products named for their beneficial effect on microorganisms, are designed to do.

200 Ways to Cut Drug Use
Today, there are some 200 probiotic products available to American farmers as gels, boluses and dry mixes for blending or topdressing. Some can be bought from seed and feed stores, while others are sold only to veterinarians or feed manufacturers.

Many, like Pioneer's Probios and Probiocin lines, contain genetic relatives of the *Lactobacillus* bacteria found in yogurt, sauerkraut, sausage and pickles. "We use the same basic culture in all of our products," says Crum. "We get them either from animals' guts or from plants. They're non-disease-causing, lactic acid-producing organisms."

Unlike antibiotics, these microbial materials do not "sterilize" the gut of an animal. In fact, they do just the opposite. They repopulate the animal's small intestine with beneficial organisms that, under normal conditions, grow there naturally. Why do that? Because most modern livestock operations offer an animal anything but normal conditions, says Crum.

Long-distance travel and crowded living quarters place today's livestock under enormous stress, he explains. That can lead to a poor appetite and a decline in the number of beneficial bacteria in its gut. "When an animal is under stress, its natural immunity goes on hold. What we try and do is provide a new source of (beneficial) bacteria that can help the animal defend itself . . . get it back on feed quickly, and keep it there."

Crum says that even though antibiotics initially wipe out all forms of intestinal microlife, sick animals treated with drugs can eventually rebuild their own defense systems. But in the time it takes to do so, the illness often causes a profit-robbing weight loss. "A feeder calf, if it gets sick, could lose 20 to 30 pounds before it gets well," he says. "You've got to put that weight back on just to break even."

Since probiotics are not drugs, they won't be effective against a disease that has already overwhelmed an animal, notes Crum. Nor will they have much effect on an animal whose gut already contains an adequate number of beneficial bacteria.

But they can be a cost-effective way of helping livestock rebuild their immune systems when they're most susceptible to disease, Crum asserts. For example, say a cattleman buys 20 calves from a distant source. Statistics show that four or five of them are sure to contract some kind of illness. Combining the cost of antibiotics (about $15 per head) with the total weight loss (about $35 per head) means that the illness cost the farmer from $200 to $250, says Crum.

"Our tests show that we can reduce the number of animals that will get sick by half," he adds. According to Pioneer's research, had the farmer treated all 20 calves with beneficial bacteria, as few as two calves would become ill. At $1.50 per head, the total cost of the probiotic treatment comes to $30. Add in the $35-per-head cost of the weight loss in the two sick animals, and the farmer would save up to $140—more than 50 percent.

Theory 'Sound'

One thing probiotics do have in common with antibiotics is that no one fully understands why and how they work. "I think it's theoretically possible," says Dr. Dwayne Savage, a University of Illinois microbiologist who works with many of the bacteria used in probiotics. "They tend to help cattle grow on less feed . . . to improve feed efficiency, and perhaps, in some cases, help them resist disease."

Agrees Lou Sudoma, product marketing manager for Christian Hansen's Laboratory in Milwaukee, Wis.: "From a theory standpoint, it's very sound. But from a technology standpoint, the probiotic industry is still in its infancy."

Christian Hansen's manufactures two different lines of probiotics for sale exclusively to feed manufacturers. Biomate FG Concentrate is a blend of *Lactobacillus acidophilus* (which is also found in Pioneer's products), and *Bacillus subtilus,* a different bacteria that Sudoma says is more capable of withstanding the high temperatures to which pelleted feeds are subjected.

While this product also helps naturally fortify animals' defense systems, Sudoma is not permitted to say so in advertisements. "It's obvious that these products are generally recommended as safe," Sudoma points out. "There are no withdrawal periods, and that's according to the FDA." But beyond that, the promotional claims that he and other probiotic manufacturers can make are severely restricted by FDA regulations. "You can say what organisms it contains and a guaranteed viability (number of live organisms). And if you've done some university studies, you may be able to show the raw data, but not make any direct claims for it. That's kind of a gray area."

At first glance, laws governing the use of antibiotics appear more lenient. "As long as the manufacturer has data to show (safety, effectiveness and withdrawl times), we can approve the product," says Dr. Max Crandall of FDA's Center for Veterinary Medicine in Rockville, Md. But in the early '70s, FDA added a new requirement that is just beginning to have a major impact on the meat industry: The makers of antibiotics must prove that bacterial restistance to their products will not be transferred to antibiotics used in human health care.

Many of the $270 million worth of antibacterials used by the livestock industry conform to this regulation. But penicillin and tetracyclines don't have to, because they were already on the market when the new law went into effect. "Once a product is on the market, we have to show that it is harmful . . . before we can get it off the market," Crandall explains.

Low, "sub-therapeutic" levels of penicillin and tetracyclines have been fed to livestock for growth promotion and disease control since the 1950s. The same products are also widely used in human medicine. And recent studies strongly suggest that antibiotic-resistance among disease-causing organisms in animals can be transferred to human beings who consume meat from those animals. The result: "super bacteria" that no amount of drug can kill. "Any time you're using antibiotics, you're going to have resistant organisms," says Crandall. "The question is: How much does the feeding of subtherapeutic doses contribute to this?"

Digestive Aids Boost Growth

If a ban on the low-level feeding of antibiotics is forthcoming, such products could, and most likely would, still be used therapeutically to fight disease outbreaks in animals. In that case, administering probiotics to a herd—like the mythical farmer in Crum's example—could keep drug costs to a minimum.

But would farmers be without the growth-promoting effects of the products they're using now? Not at all, say Crum and Sudoma. Many of the bacteria in probiotics serve as digestive aids. For example, the *B. subtilus* in Christian Hansen's Biomate FG Concentrate and Biomate 2B "breaks down protein into the readily absorbable amino acids," says Sudoma. "It takes some of the pressure off the animal and aids in the digestive process." Ultimately, that leads to improved feed conversion, he says, quickly adding that nobody is sure why this happens.

Some companies specialize in products designed solely to "predigest" high-fiber feeds into sugars and starches, much like those which break down lactose in milk for humans who can't tolerate this complex sugar. "We manufacture generic enzyme products for many, many feed companies," says Jim Tobey, vice president and director of research for G.A. Jeffreys & Co. Inc. of Salem, Va. "I know that many of them use them in feed, and many formulate them in silage fermentation aids."

G.A. Jeffreys' products are not probiotics by definition, Tobey points out. "What we're talking about here are feed items. They're really used more as processing aids than for growth promotion."

Diamond V Mills of Cedar Rapids, Iowa, produces a dry yeast mixture that farmers can blend with dry feed or a silage-grain ration. Says Dr. Charles Stone, the company's director of technical service and nutrition: "It does a lot of different things. In dairy cows, it stimulates the number of fiber-digesting bacteria. It also provides unidentified growth factors. Most of our research indicated that it stimulates bacteria to improve digestion."

Currently, only a small percentage of feeds are treated with enzymes or yeasts to improve digestion, says Tobey. That's because the sugars and starches in corn and soybeans are more easily metabolized than those in high-fiber materials like oats and barley.

In the long run, though, Tobey feels such products can help the livestock industry—and agriculture in general—become less dependent on corn and soybeans. "We can make our food chain more efficient by feeding the high-fiber feeds to the ruminants that can digest them, and leaving the more readily available foods for humans (and other non-ruminants)," he explains.

Whether it's for disease-prevention, growth-promotion, or both, livestock experts agree that benign microbial products are here to stay. "There's a lot of interest out there, and I think that interest is going to increase," predicts Sudoma of Christian Hansen's Laboratory. "There's always going to be a place for antibiotics, but I do think there's going to be a ban on the broad-spectrum products (like penicillin and tetracyclines)."

University of Illinois' Savage takes and even broader view: "In the end, what it boils down to is that neither antibiotics nor probiotics are going to be substitutes for good animal husbandry."

Building A Better Fly Trap

If you don't like flies (and who does?), here's a safe way to have a fly-free zone.

RICHARD A. FORD

COLLINS, Mo.—For years, I built fly traps using oil burner transformers and blue lights. But I didn't like the idea of being tied to an extension cord any more than the fact that commercial traps with only about $15 worth of parts retailed for around $100.

So, I unplugged the extension cord and built an enlarged version of a wire cone fly trap that was popular—and effective—during the early part of the century. The trap uses no poisons or electricity, but it sure does catch flies. It costs only $16 to $20. Baits of watermelon rinds, "supersweet" pig pellets, fish, liver, sugar syrup or separated milk lure flies into a screened enclosure where the flies become imprisoned and die.

At the end of a good week in September, a two-inch blanket of dead insects covered the base of my trap. That included a few horseflies, some wasps and moths—and about 15,000 dead flies.

Although my trap was enclosed with aluminum wire screen, using plastic sheeting, glass or close-knit cloth, such as muslin, would work equally well. The trap consists of a wooden frame, 15 inches square and 32 inches tall. The four sides and the hinged top are covered with screen. The bottom is made of a 15-inch-square piece of half-inch plywood with a 10-inch hole cut in the center. Over the hole is fixed a screen cone that extends up inside of the trap to within a few inches of the top. The trap box sits on a standard hog feeding pan, which contains the bait.

The trap box covers most of the bait pan. When flies try to leave the pan, they usually fly upward, exit through a small hole in the top of the cone and are trapped inside the

screened box. The summer heat, breezes and vibrations from the trapped flies' wings lure more and more flies to the trap.

Place the trap out of reach of farm animals and anchor it to prevent wind damage. I just ran wires from nails in the base of the trap to wooden legs holding the trap a few feet off the ground.

The trap should be about 100 feet upwind from the area you want to keep free of flies. Keeping the bait moist helps speed its decay. The bait usually has to be replaced once a week. That's a good time to unlatch the top and empty out the sun-dried flies. They make great fish food or compost.

When building the box, eliminate all cracks of more than 1/16-inch, since flies can squeeze through such small openings. The four rectangular side frames are made from 1-by-2-inch lumber. Pre-drill, glue and nail each of the four corners of each frame, then join the frames by the same procedure.

Protect the box, inside and out, with a coat of linseed oil or yellow enamel. Flies have an eye for these colors. A glossy paint also helps in cleaning out the box. Although screen can be secured to the sides with staples, a tighter box can be made by using 3/16-inch-wide wood cover strips ripped from 1-inch lumber with a table or radial arm saw. The strips can be attached over the screens with 3/4-inch wire brads. Some expense can be saved by recycling wooden storm window or screen frames.

The top is made from 1-by-2-inch lumber with pre-drilling, gluing and nailing. Three hinges are used to attach it to the top of the box. Be careful to leave no cracks between the top and the box. The latch on the opposite side of the lid should hold the lid tightly.

After cutting the hole in the baseboard, form an aluminum screen cone that is 10 inches in diameter at the bottom, 3/8-inch wide at the top and 29 inches tall. Be sure to leave a 1/2-inch flange around the base of the cone so that the cone can be stapled or glued to the baseboard. The seam of the cone can be a simple lap, secured by weaving carpet thread or a strand of 30-gauge copper wire the length of the cone. The base is then glued or nailed to the bottom of the trap box.

The main object of the trap design is to allow flies below the cone to see light coming from above. The use of the trap can be extended into cooler weather by using plastic or glass glazing to trap the sun's heat, which also helps draw flies.

My trap was based on a plan first mentioned in *Popular Mechanics* (June, 1923, page 969) by Florence L. Clark, a McGregor, Iowa, dairyman who said the trap made his milking easier and increased his milk production.

Materials list:

Recommended lumber:
  #2 white pine, yellow pine, fir or spruce
Two frames:
  1 × 2 × 12" and 1 × 2 × 30¾"
  (two each per frame)
Top lid:
  1 × 2 × 10½" and 1 × 2 × 30¾"
  (two each)
Base:
  ½ × 15 × 15" exterior plywood
  (one each)
Cover strips:
  3/16 × 3/4 × 13" (eight each)/box sides
  3/16 × 3/4 × 30" (eight each)
  3/16 × 3/4 × 13" (four each)/lid
Nails:
  6 penny galv. finish nails (40 each)
  10 penny galv. finish nails (40 each)
  3/4" × 18-gauge wire brads (small box)
Hinge:
  1½" open × 1½" long (three each)
Latch:
  small hook and eye (one each)
Feed pan:
  standard hog feeding pan 16" to 17" outside diameter
  (one each)
Aluminum screen:
  14 × 30" (four each) sides
  14 × 14" (one each) top
  32 × 32" (one each) cone
Paint, exterior enamel:
  bright yellow or orange (one pint)
Copper weaving wire:
  30-gauge × 4 feet (small roll)
Waterproof wood glue:
  2 oz. tube. (one each)
Estimated cost:
  $16 to $20 (all new materials)

Richard A. Ford is a machinist. He was living on his parents 210-acre grain and hog farm in Collins, Mo., when he wrote this article.

How To Choose: Zero In On Insect Pests

 Pest Control BioSelector™

Use our **BioSelector™ Chart** to select the best safe product to halt any of 40 common insect pests. Controls are listed left to right by their gentleness to the environment.

Variety selection	Planting date	Crop rotation	Trapping crops	Pheromone traps	Catch traps	Common Insect Pests	Bt - B. thuringiensis	Other pathogens	Dormant oil	Safer's soap	Miscible oil	DE-Diatom. Earth	Parasites	Predators	Ryania	Sabadilla	Rotenone	Pyrethrins
						Deterrent Actions / Traps	**Biologicals**		**Minerals, oils**				**Beneficials**		**Botanicals**			
						Alfalfa Caterpillar	★					★	★					
					★	Alfalfa Looper	★					★	★					
★		★			★	Aphids				★			★			★		★
						Asparagus Beetle						★		★			★	
						Bagworm	★											
★	★					Cabbage Butterfly	★						★					
★	★			★		Cabbage Looper	★						★			★	★	★
				★	★	Cockroaches						★					★	★
				★	★	Codling Moth	★	★	★		★		★	★	★	★		★
★	★		★	★	★	Corn Earworm			★		★		★	★	★			★
★		★				Diamondback Moth	★							★				
★	★	★		★		European Corn Borer							★		★	★	★	★
						Fall Webworm	★											
						Flea Beetles						★				★	★	★
						Fleas			★		★						★	★
★						Grasshoppers		★						★	★			
						Green Cloverworm	★											
				★		Gypsy Moth	★							★				★
						Hornworm	★						★	★				
				★	★	Housefly							★	★			★	★
		★		★		Japanese Beetle		★									★	★
					★	Leafhopper				★		★			★			
				★		Leafroller	★						★	★				
						Mealybug			★	★	★		★					
★	★		★			Mexican Bean Beetle							★				★	★
★						Mites			★	★	★		★				★	★
				★		Oriental Fruit Moth					★		★	★	★			★
						Pear Psylla			★	★	★		★					
						Scale			★	★	★		★					
★	★	★	★			Spotted Cucum. Beetle									★	★	★	
				★		Spruce Budworm	★											★
	★					Stinkbugs					★				★	★	★	
★	★	★	★			Striped Cucum. Beetle									★	★	★	
						Tent Caterpillar	★											★
						Thrips				★			★		★			★
				★		Tobacco Budworm	★						★	★				★
				★		Tomato Fruitworm	★						★	★				★
						Tussock Moth	★											★
						Velvetbean Caterpillar	★											
					★	Whitefly				★		★						★

© Copyright 1982 and 1987 Necessary Trading Co., New Castle, VA 24127. All rights reserved. BioSelector™ and Necessary® are trademarks of Necessary Trading Co. For a complete copy of the Necessary BioSelector write to Necessary, New Castle, VA 24127.

Reproduced with permission.

Other Fly Traps

A pyramid-shaped trap designed by USDA researchers can help control stable- and houseflies, and even reduce face-fly populations to some degree.

The device was originally used in the late '70s for monitoring face-fly populations in pasture. It consists of three, 20- by 30-inch wooden triangles painted with white, latex paint and coated with a sticky substance called "Sticky Stuff," available from Olson Products, P.O. Box 1043, Medina, Ohio 44258.

Mounted on steel posts 3 feet off the ground, the triangles reflect light at just the right angle to attract flies.

"There was some indication that it reduced populations, but it required so much labor and was only marginally effective," says Dr. Richard Miller of the USDA-ARS Agricultural Research Center in Beltsville, Md.

One problem is that a farmer would need roughly three traps for every 10 acres of pasture, and would have to protect the structures from curious cattle. Also, the sticky goo needs to be reapplied about twice a week—possibly more often if face flies begin migrating from neighboring farms where the traps aren't being used.

Instead, Miller is testing the device against stableflies and houseflies. "We're in the process of analyzing data, but we have preliminary results showing that housefly populations are reduced around barns," he says.

Guidelines for building and using the trap are available free from: Dr. Richard Miller, Bldg. 177A, BARC-East, Beltsville, Md 20705. Phone: (301) 344-2475.

For fly control devices that are less labor-intensive, but oh-so-effective, check out the variety of Silva Fly Paper Traps sold by Silva Enviro-Control Inc., 35 Post Rd. West, P.O. Box 5091, Westport, Conn. 06881.

Bug-Killing Cover Crops
Well-managed legumes and grasses are replacing ineffective pesticides in this orchard.

ALBANY, Ga.—Despite spending $40 to $50 per acre on insecticides, pecan growers like Frank Wetherbee say aphids are becoming increasingly difficult to control. So when USDA Entomologist Walker L. Tedders asked Wetherbee to test a new pest-control idea, the desperate farmer didn't hesitate.

He fall-planted alternating, 10-foot strips of hairy vetch and crimson and arrowleaf clovers between the pecan trees on all 1,200 of his acres. "We had to," recalls Wetherbee. "We ran out of chemicals."

That was in 1984. In 1986, Wetherbee didn't even spray once for aphids. In fact, he didn't spray insecticides at all until September—for leafhoppers. And his yields were just fine.

The reason for his success: A growing population of lady beetles, wasps and lacewings now make their home in Wetherbee's cover crops, and feed on aphids and other pests. "When the legumes bloom and die out in late spring, (beneficial insects) move up into the trees and clean those aphids up," says entomologist Tedders, who's been studying biological pest control for 10 years at the Southeastern Fruit and Nut Tree Research Laboratory in Byron, Ga.

The predators control aphids through August, when summer's heat causes a natural collapse in aphid numbers, he adds. And as the naturally reseeding legumes regrow, Tedders anticipates similar control in future years.

Cover-cropped apples yielded three times as much fruit as those in a clean-disked orchard, and twice as much as those in conventionally sprayed orchards, according to a California study.

50% More Predators

Cover crops don't guarantee such results overnight, warns Dr. Miguel Altieri, a biological pest control specialist at the University of California, Berkeley. "It takes time before you notice a difference," he says.

In his area, apple growers may spend up to $60 or more per acre on pesticides for codling moth, alone. The sprays reduce fruit damage to just 1 percent from codling moth larvae, and yields usually top 360 apples per tree.

For three years in the mid-80s, Altieri compared insect counts in adjacent, northern California apple orchards where conventional insecticides were replaced with sulfur, insecticidal soaps and summer oil sprays. The floor of one orchard was disked twice (spring and fall), while the other was seeded to a cover crop.

After two years, fruit damage was about 40 percent in the clean-disked orchard—about average for organic production, Altieri says. But fruit damage was just 4 percent in an orchard cover cropped with 10 percent bell beans and 90 percent mixed grasses. Altieri suspects that's because predator populations were 50 percent higher in the cover-cropped site.

First-year production in the cover-cropped orchard was below that of the disked one—214 apples per tree compared with 260—perhaps because of competition and moisture stress, says Altieri. Local orchards treated with conventional insecticides would have yielded about 30 percent to 50 percent more than the disked one. But with insecticides sometimes accounting for half of an apple orchard's production costs, Altieri says a farmer who replaced chemicals with cover crops would be more than compensated for the initial yield loss.

Yields rebounded significantly during the second year of Altieri's study. It was a dry year, and the cover crop was mowed in early April, with the residue left as a mulch. Cover-cropped trees produced an average of 334 apples each, more than three times as much fruit as those in the disked orchard, and twice as much as those in nearby, conventionally sprayed orchards.

Why did the cover-cropped trees perform so well? "Probably because of the cumulative effect of two years' nitrogen (from legumes), and the mulch's effect," says Altieri. "But these things still need to be studied."

Timely Mowing Helps

Altieri plans to continue analyzing data and fine-tuning management methods farmers can use with cover crops. Meantime, he says farmers who want to try the idea should either establish the cover crops before tree leafing, or seed shade-tolerant species.

They should also manage the covers primarily as pest control. That means carefully monitoring pest and predator populations, and mowing, if necessary, to encourage predators to feed on pests in trees rather than those on the ground.

Flowering cover crops appear to attract the most predators, so mowing when the plants are in full bloom is most effective at driving predators up into the trees. In one orchard Altieri studied, Ladino and strawberry clovers grew and bloomed throughout the season, attracting the most predators.

Still, most any cover will attract more predators than you'd find on a bare orchard floor, he adds.

Farmers in the South who adopt Tedders' choice of winter annuals may not even have to mow. Hairy vetch, crimson and arrowleaf clovers bloom and die out in late spring, just when pecan trees need the most relief from aphids. The trees also benefit "from a nice kick of nitrogen at a good time," says Tedders. "Other growers are picking up on this."

Unfortunately, few, if any, of those farmers are really cashing in on the legume N by plowing or disking the vegetation, he adds. That's because of the way pecans are harvested. Each November, farmers mow, rake and haul away the vegetation before they harvest mechanically by shaking the trees. Otherwise, "you won't be able to find the nuts," explains Tedders.

But thanks to the early-season aphid control provided by Wetherbee's cover crop, there will at least be a crop to harvest—and a lot fewer chemical bills to pay.

Sticky Snakes!

Your customers will really appreciate this neat pest control trick.

American ingenuity is not dead. In some cases, problems can still be solved in simple ways. For instance, Dr. James E. Knight, an Extension wildlife specialist at New Mexico State University, has developed a new method of removing rattlesnakes — and other snakes — from human dwellings.

"In my part of the West, people are plagued with rattlesnakes around, or worse yet, in their houses. In the past, scientists and researchers had no method they would offer when they received frantic calls for help," Knight says. "Desperate individuals spent much money and effort on ineffective and sometimes dangerous methods to get the snakes out of basements, crawl spaces, refrigerators or cupboards."

So Knight developed a safe way to rid homes of rattlers. The glueboards used by rodent exterminators are the answer, Knight found. Glueboards are pieces of cardboard coated with a sticky substance similar to that used on fly paper.

Knight capitalizes on a peculiar behavior characteristic of snakes when he places these boards in homes. "Snakes like to crawl along things. They go along sides of walls and rocks. They rarely go across open spaces. For this reason, the glueboards are tacked on thin sheets of plywood and placed tightly against the wall inside or under the house or in a basement," Knight explains. "As the snakes move around, they follow the walls, touch the glueboards and become stuck fast."

A glueboard less than one square foot in size can hold the biggest snakes, because of the large surface area of their body that becomes stuck. Knight says the amazing thing about this method is that, so far, tests have proven it 100-percent effective. One successful effort took place at a home in Edgewood, N.M. "John Morton of Edgewood lives in an area plagued with rattlesnakes. Although the family kills dozens of snakes every year, when his wife discovered a rattlesnake in the kitchen while their 2-year-old son was playing nearby, they became frantic," Knight says.

So the wildlife specialist was called in to try his glueboards. Morton was first instructed to clear the area around the house of all possible hiding places for snakes and to plug all visible holes in the cellar walls to prevent more snakes from entering. Glueboards were then placed in three areas in the basement.

"Over a two-month period of time, 25 rattlesnakes, two bull snakes and 16 garter snakes were removed from the cellar," Knight said. "Mr. Morton believes his snake problems are now over. He and his family have not seen a snake in the house for over a month."

Knight said the simplicity of the method is one reason it is so effective. The glueboards, which are available from some exterminators for less than $1, are tacked on a piece of quarter-inch plywood about 26 by 16 inches in size.

It is important to avoid attaching or placing the board near anything the snake can use for leverage that might allow it to exert enough pressure to overcome the glue, Knight adds. But none of the snakes captured by Knight have escaped the glueboards, including a 5½-foot diamondback rattlesnake.

A small hole should be made in the plywood to allow retrieval of the board with a nail in the end of a long stick.

"After the snake is captured, it can be disposed of, or vegetable oil can be poured on the glue to break it down and the snake can be released unharmed," says Knight.

Put The Bite On Snapping Turtles

JOHN M. MULLIN

Game farms that produce waterfowl are particularly concerned about snapping turtles. The snapper can be a terrible predator on young waterfowl.

Even in regular ponds, snapping turtles should be controlled because they have no natural enemies. Also, snapping turtle is a gourmet delicacy — among the best-eating of reptiles.

Many game farms harvest snapping turtles with the use of throw lines or jug lines. Single or treble hooks are baited with a tough piece of meat (neckmeat of beef is often tough enough). Be sure there is a strong swivel and a length of wire leader above the hook.

Some people claim they can catch snappers by hand in open water, especially if the pond is green with algae. They

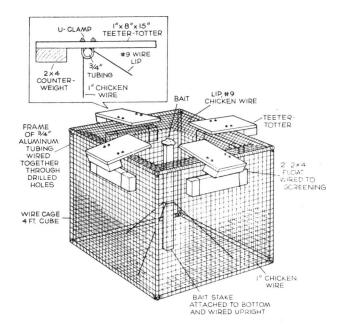

approach cautiously and freeze when they spot the turtle's head. When the turtle starts to dive, they quickly grab the turtle by the tail. I've never caught turtles this way, but I have removed snappers by the tail from seines or hoop-nets. Snappers are a handful, weighing anywhere from six to 22 pounds. The average snapping turtle will weigh about 12 pounds.

Unlike other native turtles, the snapper cannot pull its head, legs and tail into its shell.

In some nets or traps, the turtle enters through an opening in the bottom. The diagram at left shows a "teeter-totter" trap where the turtle enters from the top. The turtle crawls out on the teeter-totter board just above the water. Its weight causes the teeter-totter board to tilt and the turtle slides into the trap.

This teeter-totter turtle trap involves complicated construction, but properly built, it is long-lasting and effective.

I've never found any really easy way of processing a turtle. But I know that most edible turtle meat is on the legs, neck and two tenderloins in the back. These cuts can be removed and prepared for some delicious eating. Soak the meat in cold saltwater. Then mix about half a cup of flour, a little salt and pepper in a paper bag, drop in a few pieces of turtle meat and shake until the meat is dredged with the seasoned flour. Add half a cup of oil to a medium frying pan and brown the meat rapidly on both sides. Then, reduce heat, add half a cup of water or whatever liquid you wish to add moisture and cover the pan. Cook slowly until the meat is tender.

John M. Mullin is editor of Wildlife Harvest, *Goose Lake, Iowa, 52750.*

Part III
The Man

Chapter 10
Conversations With Booker Whatley

Editor's Note: *The following question-and-answer session was taped during a 10-hour trek from Boston to Emmaus, Pa., on Nov. 20, 1986, after Booker Whatley spoke at the Conference on Sustaining Agriculture Near Cities. Ordinarily, that's only a 6-hour drive. The delays included an ice storm that caused a massive traffic jam on I-84 in Connecticut and New York, heavy fog, and an iced-up fuel line that almost left Whatley and New Farm Editor George DeVault stranded in the pouring rain in the boondocks of the Delaware Water Gap National Recreation area at night.*

Q: *Can you really make $100,000 farming 25 acres, I mean legally?*

A: You can make a whole lot more than that, providing you're a good manager and really use your head. I'd say that one of these farms, properly run, could gross in the neighborhood of a quarter of a million dollars a year after it comes into full production.

Q: *How long would it take to reach that level?*

A: Oh, to get everything the way it ought to be, I'd say it would take most people three to five years. Probably closer to five, because that's about how long it takes before your blueberries, kiwi, grapes and other crops like that start bearing very heavily. That's why, when you're setting up a farm like this, you want to get those crops in the ground right away. It takes time for them to become established. A lot of people don't realize that. They just hear "$100,000 a year" and think they can plop a few seeds in the ground and start raking in the money before long. It just doesn't work that way. If it was that easy, why, then everybody would be doing this.

Sgt. Booker T. Whatley, the oldest of 12 children, with the youngest Whatley, his brother Moses. (circa 1942)

Q: *How many farmers are following your plan?*

A: I can't say, for sure. There are about 1,000 that I know of, but I only know about them because they are all subscribers to my *Small Farm Technical Newsletter*. There are probably a lot more that I don't know about.

Q: *How are they doing economically? Have they hit that $100,000 mark?*

A: Well now, I haven't personally gone over all of their tax returns, you understand. So I couldn't tell you exactly how much money they're making. But I'd say the majority of them are doing just fine, at least the ones who have been at it for awhile are. Why take Frank Randle, for instance. You met him when you were down in Montgomery for my birthday a couple weeks ago. (See the article "Being Profitable Means Being Different" in Chapter 5.) Now that boy is making some good money off his little place, and his only cash crops are blueberries, honey and lamb.

Q: *And Frank doesn't even have a Clientele Membership Club. At least he doesn't charge his regular customers an annual membership fee. How many other farmers are following just part of your plan?*

Maj. Whatley during the Korean War

A: Let me answer this way: Nobody that I know of is following my concept and plan 100 percent. I don't think there ever will be. Americans are independent and farmers are even more independent than most of us. We're a free country and I hope we stay like that. If everybody started doing just what I say, it would mean they aren't using their heads enough and coming up with enough ideas of their own. They'd be relying too much on someone else's formula or recipe. That's just what got farmers into a lot of the trouble they're in today.

Q: *OK, but how do you answer the critics who try to shoot down your idea by saying it's "scientifically untested," that you never actually grossed $100,000 a year from a commercial farming venture?*

A: Folks who automatically criticize an idea just because it hasn't been randomized and replicated, like they teach at all the land grant colleges, are walking around with their eyes closed. They just don't want to see the possibilities for other ways of doing things. People who say that I haven't "proven" my concept . . . and, you're right, there are a whole bunch of them . . . they are missing the point. When you're dealing with a concept, you don't have to demonstrate it. All you have to do is explain it. Other people will take it from there. They'll refine it, modify it and do whatever they need to do to make it fit their peculiar needs and circumstances.

Q: *You didn't have too many kind things to say about the land grants and USDA, today. Why?*

We're All Workers In The Same Vineyard

Our two youngest nieces, Sham and Tam, are twins. When they were 8 years old, they wanted to start a social club for themselves and 10 of their 8-year-old friends. They phoned their friends and explained what they had in mind. The girls accepted. The first meeting was set for a Saturday afternoon.

Meantime, they had me take them to the supermarket and pay for what they wanted to serve. The date finally arrived for the meeting. Less than 20 minutes into the meeting, it became deadlocked. Each of the 12 members wanted to be president.

I sat there in total astonishment as someone made a motion that the president could serve for one month or one meeting only. The motion was seconded and carried.

Those girls are now 16. Their club still meets every month. And each member has been president eight times. We adults and small farmers can surely learn from those little girls. First, we must determine what the problem is. Second, we must take action to solve it.

The other lesson to be learned is that you can solve your own problems. What those 8-year-old kids did was eliminate the big "I" and little "u," and place all members on equal footing. Small farmers must accomplish this with respect to their Clientele Membership Club. In the sight of God, we are all equal.

Dr. Whatley upon receiving his doctorate in horticulture from Rutgers University in 1957.

A: Because we . . . and I can and do say "we", because I was educated and trained in, and spent years working in the land grant system . . . we agricultural professional workers have failed our farmers. We have failed them, miserably. That's especially true with our small farmers. We told them 25 or 30 years ago to get the heck out of the way, because they just didn't matter. Then we went ahead and fed the rest of our farmers such a load of garbage that many of them are about bankrupt today.

Q: *You were saying over lunch that we're seeing some fundamental changes in agriculture that historians are going to rank right up there with the invention of John Deere's plow, the tractor and hybrid seed corn. How about looking into your guru's crystal ball and telling me what you see happening in farming by the turn of the century?*

A: We have to stop thinking that subsidies are the solution to our farm problems, because they're not. God knows, if money had been the solution, we would have had it solved a long time ago. Money is not the solution!

At least not money from the government. What the farmer needs and is going to have to have is a greater share of the housewife's food dollar. He just has to have more. He can't make ends meet the way things are. All these middlemen, they're a greedy bunch. They just take it all! Also, a lot of farmers can't switch to this diversified farming because they've lost all their equity. They just don't have any equity left.

Q: *What happens then?*

A: I think we're going to see most of those guys go out of business and a lot of new people come in. The big boys can't make it. I don't care how big you get. How long can you last when you're selling corn for 98 cents a bushel?

Now I wasn't joking. It may sound like a joke, but I wasn't joking when I said the thing that worries me is our government trying its hand at farming. They told the little guys that they couldn't make it. Once the government decides that the big boys can't make it then they're going to try their hand at it. And that's when things are going to get fouled up for sure.

Something has to give on this thing, because, face it, we just have to have our farmers. We have to have wheat, corn and soybeans. We have to have dairy, beef cattle and hogs. We have to have all that stuff. But how can you do it under the present setup? You can't do it.

Q: *Of course, some people say it doesn't matter. We can buy whatever we need from other countries that grow it for less.*

A: Sure, and some people are saying, "We can't afford these damn farmers. They're getting a subsidy now amounting to $20,000 per farm. And we can't afford it." They want to eliminate our farm subsidies, or at least most of them. And I'm all for that. Many of our farm programs were set up in the middle of the Great Depression 50 years ago, and they're just not workable in today's agriculture. Programs that were originally meant to help the little guy are giving millions and millions of dollars to large operators who are just farming the government. The little guy, meantime, is not getting a thing.

We have to be careful, though, that when we cut back on agricultural subsidies, we don't wipe out most of our farmers. Sure, $30 billion a year is too much to spend on farm programs. But can you imagine how much more Americans would be spending on food if we didn't have enough farmers to meet our basic food needs? We import too much food as it is. Just imagine what things would be like if we had to rely on other countries for most of our food. Why, it would make the energy crisis and gasoline lines of the '70s look like a Sunday school picnic. If the press and the media would start making some real concerted efforts to inform the general public of the situation and point out the real problems, somebody will come up with the solutions. That, and you have to get a lot of these farmers to admit they have a problem with their present crops and the usual way of doing things. I don't know if a lot of them know that they have a problem. We receive a lot of letters from farm women, because they're the ones who generally keep the books. They know what shape they are in, but their husbands just won't listen to them. One woman said, "I've been telling my husband for 20 years that I knew there was a better way than this."

Q: *Do any Extension agents that you know of tell people to do what you're talking about?*

> ### The Real Test Of Human Behavior
>
> In 1881, when Booker T. Washington founded Tuskegee Institute (now Tuskegee University), arriving students were not given a battery of tests, as colleges and universities do today. Booker T. Washington simply asked if would-be students wanted an education, and whether they were willing to work for it.
>
> Mr. Jesse D. Mason hoboed (rode freight trains) from Arkansas to Montgomery, Ala., then walked and hitchhiked the last 40 miles to Tuskegee Institute. He arrived with 75 cents in his pocket. Affirmative answers to those two questions told almost everything about the student's character, inspiration, honesty, motivation, and determination. The battery of tests colleges and universities administer to incoming students today tells very little about those aspects of human character.

A: I don't think so. We have a few of them who subscribe to our newsletter, but the system won't allow 'em to really do much. The system is just not designed to let them do anything other than just what they were hired to do. That's why we have soybean specialists, peanut specialists or 4-H club specialists, all that kind of stuff. There is just no room for a small-fruit or game-bird guy. Extension is just not hiring those folks. And all most Extension people can do, really, is just hold meetings. Now they're *good* at holding meetings, you know. They hold a whole lot of them. And then they think that you can solve a whole lot by getting on radio and TV. Well, they hold those programs at the wrong damn time, because everybody is at work. But they still get on radio and TV.

One thing Extension is doing these days is talking about crop diversification, and that really worries me. Why? Because they're used to dealing with just a few things at a time and you can't just do this piecemeal.

It's like that ADAPT 100 conference I'm going to next month. One thing I want to put into my speech is that it's alright to have high-value crops, *but they have to be compatible.* Many farmers have not thought about that. They're not using a systems approach. The guy talking about raising alligators will just talk about alligators, and you'll see a lot of farmers jump into a lot of things that they can't afford to jump into.

If the Extension Service was doing its job, agents would be out there working night and day, advising these farmers on just how to go about all this in a systematic way so that we don't end up with the exact same problems we have right now. But they're really not doing one thing different, when you come right down to it. That's why these farmers' protests are going to spread. We're going to see more things like they had out in Manhattan, Kan., this summer. It was almost like in the '60s, when the students raised such a ruckus on college campuses all around the country. I guess it shook up Kansas State. The farmers said that school wasn't doing a thing to help them. I don't know why they jumped on Kansas State, because Kansas is not unique. All those land grant schools are just alike.

Q: *Do you think they'll ever do away with Extension, as some people are suggesting?*

A: I think so, myself. If we'd do away with them, nobody'd miss them.

As far as small farmers are concerned, Extension just isn't doing a thing for them. Absolutely nothing! Well, nothing very good anyway. I get a lot of letters from prisoners, convicts. They say, "I'm up in this federal pen and that damn USDA and Extension Service helped cause me to be here." I think I'm going to start putting that in my speech. These are all educated guys. They're no dummies. One old boy from Kentucky impressed me in particular. He wrote me a long, nice letter. He said he had a 90-acre farm and he was trying to operate that farm just like a guy in Texas who had 10 sections. And he said that when he realized he was going to lose his farm, he took up a temporary occupation — as a bank robber. That joker still impressed me. He said he liked farming and he wanted to get back in it. They sent him up to a federal pen in Jackson, Mich. When he got released, we never heard of him and the prison won't give a forwarding address.

Q: *How about Tuskegee? Are they still doing anything there with small farms?*

A: Very little. In the first place, like I told you, they don't believe in it. And they criticized me. They wanted to use the term "limited resources." Well, as far as I'm concerned, when they say limited resources, what they mean is a guy who is limited in the head, limited in the pocketbook, limited in management skills, limited in the ability to think, you know. Limited every kind of way. But all I'm talking about is a guy who has limited land. They're preaching Luther Tweeten's song. (Luther Tweeten is an agricultural economist at Oklahoma State University. He is a frequent critic of small farms and the Whatley plan.) They just sing it all the time, you know.

Q: *Did you ever meet Tweeten?*

A: Yeah, I met him up at Beltsville (the Beltsville Agricultural Research Center in Beltsville, Md., at USDA's special symposium on research for small farms, Nov. 15-18, 1981).

Q: *You didn't get in a fight did you?*

A: No, we don't ever fight. I guess we're friends. He's got a right to his opinion, just like I got a right to mine. And I realize that.

Q: *Is anybody teaching your concept as a college course? It seems like the kind of thing a really innovative professor could have a lot of fun with.*

A: Nobody that I know of. See, the way this thing is structured, in a white land grant college, they hire you to do a specific job. Now one good way for you to lose that job is to not do what you were hired to do.

In a black college like Tuskegee, they don't hire you to do anything in particular. They hire you because you have a degree. They could care less what you do. As a black professor, I used to do anything I wanted to. Nobody told me to breed sweet potatoes or honeybees. I just did it. You have all the freedom you want. But let me tell you something, doing nothing is habit forming. When you can do nothing and still get your pay, there are a whole lot of folks who are going to do absolutely nothing, I tell you.

The only thing that makes it difficult for you in a black school, is if you have a professor who is going to be innovative and show initiative. Then the other professors are going to stand around and criticize all the time, you know.

I learned a long time ago the way you survive with that bunch: You have to act just like they don't exist. That's the way you get around them. That's what I did at Tuskegee. Just don't pay them any attention. Then they're not a threat. They're not a problem, until it comes to money.

At Tuskegee, I never got any money for small farming, even though Tuskegee was getting a lot of money from USDA. The only way I ever got my money for small-farm work was from Rockefeller. And I used to get all the fertilizer I needed from TVA. They *gave* it to me. And when I got ready to build that 14-acre pond, I got the money from the computer people out in Minnesota, Control Data.

Whatley was just starting to work with small-farm cropping systems in the summer of 1974 when then-Agriculture Secretary Earl Butz toured Tuskegee Institute.

Q: *Are many black farmers adopting your plan?*

A: We have more Canadian subscribers than we have black subscribers in the state of Alabama. That's because with old blacks in my age group, they worked and sacrificed and bought that land, most of them. And they sent their kids to college. The kids got out of school and they went up North. The old folks would go off and work in town and do *anything* to get up enough money to pay the taxes on that land. But these youngsters aren't going to do that. When the old folks die, the kids sell the farms, because these farms are a damn liability. I was speaking in Montgomery at the campus of Alabama State University years ago. Man, I was up there pushing small-farm diversification and some old Baptist preacher from Huntsville came up to me after that thing was over. He said, "Doctor, the reason that blacks are not taking to your program is that they cannot conceive of a farm making $100,000. When they were growing up on their farms, they didn't get nothin' out of 'em. They just got their seeds. If I was you, I'd cut that down to $10,000. You would probably attract more." I laughed. I said, that crazy preacher! I thought about it, but I still couldn't agree with him.

When I was out at Kansas City last week, there was a young black from Cleveland, Ohio. His granddaddy, who came from right down in Alabama, bought 30 acres of land, which, at that time, was on the outskirts of Cleveland. His granddaddy worked at some of the factories there and did a little truck farming on the side and this kind of thing. His daddy held onto that land. And now he and his sisters are holding onto that land. He read about my plan in *Science '84*. The next year, they decided they'd just put in 2 acres of collards and turnip greens. And that boy said that they made $20,000 off an acre and a quarter of collards and three-quarters of an acre of turnips. He said, "We didn't harvest them all! We did everything wrong. But I see now with those 30 acres of land that all of us could get rich! I realize now that one of your farms will succeed." But I don't know of another young black, anywhere, who has attempted to set up one of these things. Not one.

Q: *Was this man in Cleveland going to go ahead and implement other parts of your plan?*

A: Oh, yes. He said they were going to put in some strawberries and blueberries late this year or in the spring.

Far as I'm concerned, there's nothing like blueberries. A farmer can get rich selling blueberries for 60 cents a pound. I mean he can really get rich! Once you get that crop established, it's going to be there for 40 or 50 years. With a Clientele Membership Club, your customers do the picking and you save about 50 percent on your production costs.

I just don't agree with that fellow back in Boston who said he thought you ought to charge 40 percent more, rather than 40 percent less than supermarkets. You have to set up a system that has incentives in it for everybody. Everybody likes to save money. We don't have to get greedy and try to take it all.

Q: *Don't CMC membership fees and other things tend to make up for the discounts on U-pick crops?*

A: Sure they do, especially if you can command a higher membership fee. I usually recommend an annual membership fee of $25. But I know several women on the West Coast who set up CMCs and they charge $50. That's the most I've ever heard of anyone charging. The least I ever heard of was $5. And that didn't last very long. When this fellow was just starting to set up his farm, several women came by and said, "You're not charging enough for this. You ought to charge at least $25." His teeth like to fell out!

Q: *Now that doesn't include fishing rights, does it?*

A: No, that doesn't include fishing rights—or hunting rights. For fishing rights, I suggest that people charge an additional $15 a year. Now that just includes the right to fish in your pond. It dosn't include the fish. For every fish a customer catches, they pay extra. In Alabama, places where you catch your own catfish charge $1 a pound. We would let our CMC members catch fish for 40 percent off that price, or 60 cents per pound. Now the reason the farmer can do that at that lower price is because he's not fertilizing his pond or feeding his fish the way that $1-a-pound catfish place is. No, he has a flock of ducks that "fertilizes" the pond. That stimulates the aquatic growth and produces more food for the fish. If you're in no hurry to open your pond for fishing, just leave the fish alone for two years after you stock it. Then you never will be able to catch all of the fish out of that pond. But if your CMC members want fish right now, stock your pond, but then go out and buy some 1-pound fish and turn them loose in there.

And, believe it or not, you can have a serious predator problem with your fish pond. That airline pilot from New Hampshire told me something today that I'd never heard before. He has a pond stocked with trout on his farm. But it got so he wasn't catching anything from it, so he finally drained the pond. There wasn't a fish in there. Not one! He said the otters had eaten up all of his fish. He didn't have the pond fenced, but he says he's going to install some electric fencing before he restocks his pond.

With hunting, that's a whole different ballgame. You're appealing to an elite bunch of customers. These people have a whole lot of money, and they don't mind spending it to come out to your farm for some guaranteed good hunting.

One thing, though. You have to operate just like you do with your crops, and stick with high-value enterprises. Don't mess around with rabbit hunters. If some joker just wants to shoot rabbits, you tell him to go bother someone else. You don't have time to fool with rabbit hunters.

But if you have your own hunting preserve, you can charge people $150 to $200 apiece to join your hunting club. Deer is where you really make the money. It takes a sizable investment in land and money to set up a 200-acre deer-, squirrel- and dove-hunting preserve like I recommend, but the returns are also great. First, you can charge $150 to $200 to belong to your hunting club. For another $500 to $1,000, you can guarantee hunters a buck. People just go crazy for a buck with a good rack. Now this is a sophisticated bunch. You'll also have to include two nights lodging and food, meals and all that. (See Chapter 5.) But for that kind of money, it's worth it.

Q: *You said this man was an airline pilot. Isn't that pretty far removed from farming?*

A: That may be, but it's certainly not slowing him down any. The same is true with a lot of these other small farmers. They come from all walks of life. Besides airline pilots, there are corporate executives, insurance salesmen, factory workers, teachers, stockbrokers, doctors, newspaper and magazine writers and, of course, farmers. Probably the largest group of this new breed of farmer consists of dentists.

These people are the real entrepreneurs in America today. They are smart, hard-working and hard-thinking. They are not just playing farmer or looking for tax shelters. These people mean business. They're serious about this. And that means there are going to be some real changes in farming in the next few years!

'That's How Mama Always Did It'

We have a first cousin who married a fine young man who just loves baked ham. His first payday after getting married, he brought home a nice smoked ham and told his bride he wanted baked ham for Sunday dinner.

Saturday evening, his bride started preparing the ham. She first cut off the shank. He asked, "Why did you remove the shank?" His bride replied, "That's the way you cook a ham and, besides, my mother always removes the shank before baking a ham."

A few weeks later, while visiting his mother-in-law, he asked the older woman, "Why do you always remove the shank when you're preparing a ham for baking?" She replied, "That's the way you prepare a ham for baking . . . plus, my mother always removes the shank before baking a ham."

Several months later, he met his wife's grandmother and, for the third time, asked, "Why do you always remove the shank from a ham when you're preparing it for baking?" Grandmother replied, "Son, in the old days, I removed the shank because my baking pan was too short. But now, I don't have to, because some of my grandchildren gave me a much larger pan for my birthday several years ago."

Chapter 11
'The Ultimate Small Farm'

Page 35 is a long, long way from the front page of *The Wall Street Journal*. But the article there on Oct. 4, 1984 was still good enough to shake up a lot of folks all over the country.

"Booker T. Whatley Contends His Program Will Help Small Farmers Make Big Money," read the headline. *"For too long,"* the article began, *"the prevailing wisdom among small farmers has been to get big or get out, says Booker T. Whatley. His advice: Stay small, but get smart . . . start thinking less about big tractors and more about marketing."*

About the middle of that morning, the phone rang at my home in Montgomery, Ala. "This is Tom Monaghan, and I just read your article in *The Wall Street Journal,*" said the cheery caller at the other end of the line. I instantly recognized the name. He was the owner of the Detroit Tigers baseball team, which then was just one game away from winning the American League championship. I said I was pleased, but very surprised to hear from him.

"I don't know why you should be surprised," he replied. "All of us eat three meals a day and I haven't heard that eating is going out of style."

'Diversify & Profit'

Monaghan said he wanted to learn more about my small-farm plan, and invited me to visit him at his headquarters in Ann Arbor, Mich. It just so happened that I was already scheduled to be in Lansing, Mich., in a few months to address the Great Lakes Vegetable Growers Association. We agreed to get together then.

About 10 minutes before I was to give my morning speech, I went out into the hall to get a drink of water and was immediately greeted by Tom Monaghan, who recognized me from my picture in the newspaper. He and some of his top aides had slipped unnoticed into the auditorium. We talked briefly and then all went inside.

When I stepped up to the podium, I told the audience I was delighted and greatly honored by the presence of Mr. Thomas S. Monaghan, the owner of Domino's Pizza and the Detroit Tigers, who, by the way, had gone on to beat the San Diego Padres and win the 1984 World Series in five games. "Mr. Monaghan, would you please stand?" I said. As Tom rose, that crowd of 2,300 went absolutely wild!

Tom sat through the morning seminar, took me to lunch and stayed to listen to my afternoon presentation. Here's how he later described the conference in his autobiography "Pizza Tiger" (1986, Random House): "I thought, My gosh, wouldn't it be great to have one of these farms on Domino's Farms? It would add a real agricultural flavor to the place. Besides, Dr. Whatley could be a great adviser on crops we would raise on the farm, like green peppers and onions for use as toppings in our stores.

"I was completely captivated by this (then) 68-year-old black man. He came on like Bill Cosby imitating Uncle Remus; he had the audience in stitches, but he never cracked a smile, while making sly jokes based on farm folklore and taking pokes at the U.S. Department of Agriculture. He has a low opinion of farm bureaucracy."

The 'Get Big' Blues

You can say that again, Tom. My quarrel with USDA and the land grant college bunch is that they very seldom have a really new idea, let alone one that's going to do something good for the farmer. They all think exactly alike because they were trained to think exactly alike. They all went to the same schools, listened to the same professors. It's like a big fraternity. Everybody thinks everybody else is just great, because they all think alike.

For years, this bunch was telling farmers to get bigger. Get big or get out! Grow more corn, more beans, more wheat! Look at the mess that's gotten farmers into. Now, the same people are pushing this idea of diversification, but relying on traditional marketing outlets like farmers markets and cooperatives. That just won't work. Small farmers simply can't afford to pick, grade, wash, package and haul their produce maybe 100 miles or more to market, yet that's exactly what the so-called experts are telling them to do.

Tom Monaghan realized all that immediately. He is no dummy when it comes to marketing. Excellent marketing is what helped him build Domino's Pizza into a $2 billion-a-year business with some 3,800 stores in seven countries. One of the things he quickly realized during his first year in the pizza business was this: *It takes just as much time to make a small pizza as it does to make a large one, and it takes just as long to deliver a small pizza.* Dropping small pizzas caused an immediate 50 percent increase in his sales. Almost the same thing happened when he eliminated submarine sandwiches from his menu.

'Selling' Rural Charm

In June of '85, my wife, Lottie, and I flew to Detroit to select a site for Tom's Whatley-style farm. Two of his security people met us at Detroit Metro Airport. They collected our bags, ushered us into a limousine and drove us to the Marriott Hotel in Ann Arbor. After lunch with Tom and some of his top aides, we boarded the Domino's helicopter and flew over four possible sites for the farm he wanted to start. I was looking for the best exposures, pond sites and hardwood stands for use in shiitake mushroom production. We later toured the sites in a four-wheel drive truck.

All the sites had a lot going for them, but I picked the one on Warren Road with the old, run-down barn. Why? Because it could be developed into a more beautiful site, and these farms are supposed to be beautiful farms. All your customers should enjoy themselves at your farm. They should eagerly look forward to each and every visit. That's why you'll want to include things like a picnic area with shade trees, comfort stations, a small petting zoo for the children and a place where customers can at least get a drink of water. Think of your farm as a vacation spot or a retreat for your customers. Many come there to enjoy the country atmosphere and get away from the hustle and bustle of the city, as much as they do to pick fresh fruits and vegetables.

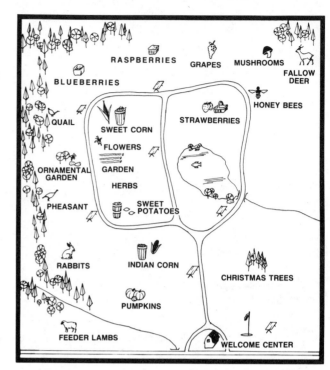

In addition to the usual high-value crops, annual production of the Booker T. Whatley Farm at Domino's Farms will include 24 tons of mushrooms, 100,000 quail, 100,000 rabbits, 50,000 pheasants, and 500 fallow deer. A flock of 100 guinea fowl will help control insects and also serve as "watch dogs." The 20-acre fishing/irrigation pond will be fertilized by a flock of 100 ducks. A renovated barn will house a classroom, visitors' center and retail store.

That night, Lottie and I were Tom's guests in his "box" at Tiger Stadium. We were joined by two very special guests who Tom had invited, Detroit Mayor Coleman Young, who I had not seen for more than 40 years, and Al Kaline, who Tom described as the greatest outfielder since Ty Cobb. Mayor Young and I were Army Air Corps cadets at Tuskegee Institute in 1942. He is especially interested in my small-farm plan because of the potential it may have for "neighborhood farms" on vacant land all around the city of Detroit. The evening was capped off by the Tigers beating Toronto, just as Lottie had predicted.

That weekend was the most enjoyable weekend of our lives. Lottie simply fell head over heels in love with Tom, and wanted to put him in her bag and take him home with her. She believes that if God ever made a perfect human being, it is Tom Monaghan.

The 'Booker T. Whatley Farm'

I went back to Ann Arbor in October 1985 and gave a seminar on my small farm plan for the Domino's family of employees. When I finished, Tom announced that a Whatley-type farm would be established at Domino's Farms, his 300-acre headquarters complex. "With your permission," he said, looking directly at me, "I want to name it the Booker T. Whatley Farm." I was stunned, and humbly accepted.

That night, I called Lottie and told her what Tom wanted to do. She was even happier than I was, if that is possible. But I had another piece of good news, too. Tom had accepted my invitation to visit us in Montgomery for my 70th birthday the next month. She didn't believe me. "Now I'm sure that a busy man like Tom Monaghan has a lot more important things to do than come to Montgomery for your birthday, Booker!" she declared. I told her that Tom had no reason under the sun to lie to me. I have faith in Tom Monaghan. As far as I'm concerned, when he tells me something, that's gospel.

Sure enough, on Nov. 5 at about 10 a.m., Tom and his party arrived at our home at 7 S. Haardt Dr. in Montgomery. There were some 30 other guests altogether. We had invited the press, too, and Tom and I sat down for a news conference. The reporters zeroed in on Tom, pressing him for views on drug use by professional athletes. He replied that he thought drugs did not belong in sports or any other part of life. Tom then turned the discussion to farming. "Dr. Whatley is trying to save 100,000 small farms in this country . . . and I am going to help him!" he declared.

Why would the president of an international food company take such a keen and personal interest in farming? There are many reasons, starting with the love of farms and farm life that Tom developed while living at rural foster homes, when not at St. Joseph's Home for Boys, a Catholic orphanage in Jackson, Mich., where Tom spent much of his youth. One of his favorite books from boyhood is Louis Bromfield's "Pleasant Valley."

Tom had been toying with the idea of having his corporate headquarters on a farm since the late '60s. His infatuation with farming also is a natural extension of Tom's being a scholar and disciple of architect Frank Lloyd Wright. In fact, Tom has probably the largest private collection of Wright art, furniture and memorabilia in the world. He is a firm believer in Wright's philosophy of "organic architecture," in which a building accommodates its setting, not the other way around. That's why Prairie House, Domino's headquarters building, was designed to follow the level horizon of the Michigan landscape. It is only four stories tall. But, when completed, Prairie House will be about a half-mile long.

(If you're planning to travel near Ann Arbor, be sure to stop for a tour of the Domino's Pizza headquarters and Domino's Farms. The address is Domino's Farms, 24 Frank Lloyd Wright Dr., Box 874, Ann Arbor, Mich. 48106. Phone: (313) 995-4500.)

Wright's position on skyscrapers was equally unconventional: "A tall building or high-rise should not be thrust into a crowded city . . . but rather should stand free, preferably in semi-rural surroundings." Next to the Prairie House, on the other side of a small lake, Tom plans to build the Golden Beacon, a skyscraper that Wright designed in 1956 but never built. Although it has only 30 floors, the Golden Beacon—with its 16-foot ceilings—will be as tall as a 50-story building.

Above all, though, is the fact that Domino's Pizza is a large food company that uses tremendous amounts of agricultural commodities. And food comes from farms. The 189 million pizzas Domino's sold in 1986 required the following quantities of ingredients:

- Flour — 125 million pounds
- Cheese — 86.7 million pounds
- Tomatoes — 1.75 million cases (6 #10 cans per case)
- Green peppers — 3.9 million pounds
- Onions — 3.9 million pounds
- Mushrooms — 260,000 cases, canned
- Green olives — 44,000 cases
- Black olives — 145,000 cases
- Ham — 4.5 million pounds
- Sausage (pork) — 9.3 million pounds
- Pepperoni — 17.4 million pounds

Now we didn't eat anywhere near that much for my birthday dinner, but we did eat our fair share of baked Pacific salmon, honey baked pheasant from Minnesota, baked country ham from Georgia, chicken lobster from Maine, wild rice, tossed salad, hot rolls and red velvet cake.

Whatley and Monaghan launched their small farm venture with a "country fair," complete with hayrides, lots of fried chicken, a horse pulling contest, straw hats and bib overalls.

We also had quite an exchange of presents. Lottie and I presented Tom with a bushel of pecans from the old trees in our backyard and a bushel of Alabama sweet potatoes. They were "Carver" sweet potatoes, one of the five sweet potato varieties I developed. I named this variety after George Washington Carver who worked at Tuskegee until his death in 1943. For some reason, there had never been a plant cultivar named after Dr. Carver. There were a whole lot of schools and streets, a postage stamp and even a sumbarine bearing his name. But no plants! And yet plants were his life. He did a lot of work with sweet potatoes.

Tom also received several bottles of outstanding homemade wine from my friends, John and Bernice King, Edward and Ernestine Stevens and Dr. J.H.M. and Bettie Henderson, and a bushel of Alabama peanuts from Dr. and Mrs. R.D. Rouse of Auburn University. Tom, in turn, presented me with a 1984 World Series bat. Only 200 of those bats were made before they destroyed the die.

Tom Monaghan is the most gracious, amazing and understanding individual I have ever met. I recommend his book, "Pizza Tiger," as required reading for every high school and college student and adult in the country. Having met and become friends with Tom Monaghan has changed my life and Lottie's. Reading the amazing story of Tom's life will give you a much more positive perspective on life. I think I like Tom so much because he and I are similar type individuals. We both can visualize something that does not exist and then develop the procedures for having it come to pass. We both have obsessions against middlemen, especially if we think we can and should perform a task ourselves. Both of us have worked hard and used our God-given intellect to get where we are today.

Breaking New Ground

Despite Tom's grand plans for the Booker T. Whatley Farm at Domino's Farms, the project stayed on hold through all of 1986. Domino's Pizza was expanding rapidly, adding more than 1,000 stores. All available funds were channeled into the expansion effort.

By summer 1987, though, construction began at the site I had picked out on Warren Road.

The official dedication was set for 4 p.m. on June 23. The evening before, Tom hosted a reception for Lottie and me, several dozen relatives, friends and various officials from Domino's and the community in his Prairie House office.

Now let me take just a minute to tell you about Tom Monaghan's office: It's a good 100 feet long—running the whole width of Prairie House—on both the third and fourth floors, with wrap-around balconies on both levels. I'd say it's 50 feet wide.

Everything is done in earth-toned brick and terra-cotta, natural wood and brass. The lower floor is more for entertaining. It has a deeply sunken conversation pit that is

Domino's Pizza President Tom Monaghan welcomes a crowd of more than 100 to the dedication of the Booker T. Whatley Farm at Domino's Farms. His new overalls, straw hat and a bandanna earned him a special award from co-workers; the "Worst-Dressed Farmer" of the day.

bigger than a whole lot of living rooms. The interior brick wall is dominated by a large fireplace. The outside walls are glass, and offer sweeping views of the surrounding countryside. A round, green-topped conference table fills the other end of the room. That table is 15 feet across if it's an inch. The area above the table is open, clear to the ceiling of the next level. A wide, U-shaped stairway, overhung with plants, leads to Tom's well-stocked library and "working" office on the upper level.

Everyone, but especially my 16-year-old twin nieces, Tam and Sham, had a good old time exploring that office. The hors d'oeuvres, made up of spiced shrimp and pineapple, chicken, fruit, cheese, mushrooms and vegetables, were as elegant as the surroundings. What really tickled me, though, was the main course, Domino's pizzas—some with everything but anchovies—served on silver platters.

The next day, this new farm was officially dedicated, with the raising of the flags of Domino's Pizza, Whatley Farms and Domino's Farms, and the unveiling of a life-sized, basswood statue of me that Tom had commissioned by Bert Rackotz, a local artist.

The other festivities included some things that you might want to copy in promoting your own farm and Clientele Membership Club:

• Ceremonial barn raising, in which Tom and I raised a small section of barn siding that had been hinged for the occasion.
• Horse pulling contest.
• Pig racing.
• Hay rides.
• Displaying old farm tractors and trucks.
• Name-the-pony contest.

'THE ULTIMATE SMALL FARM'

Lottie Whatley officially unveils the life-sized basswood statue of her husband, who is standing to the left with Tom Monaghan. The carving will serve as a mold for a bronze statue that will stand at the entrance to the Booker T. Whatley Farm.

By the end of its fifth year, the farm will be grossing more than $2.5 million a year from the sale of everything from blueberries, honey and lambs to Christmas trees, venison and catch-your-own fish. The farm will supply mushrooms to Domino's Ann Arbor commissary, which supplies the ingredients for more than 1 million pizzas a month in Michigan and parts of Ohio and Indiana. Domino's has 27 such commissaries around the country. But the farm's main customers will be the Clientele Membership Club made up of the employees at Domino's headquarters, other Prairie House business tenants and Ann Arbor area residents.

"I'm also interested in what happens here because we're trying to do a similar thing in Honduras," Tom said at the farm dedication. "Honduras is the poorest country in Central America. They just don't know much about farming. We hope to buy the land, set up the farms, the houses, show them how to do it and give them a mortgage and put them in business. We hope we can do a lot to turn that country around."

Tom has long supported the work of a Catholic priest in Honduras. What he plans to do is buy a state of the art tomato processing plant in that country. The plant comes with 8,000 acres. Tom only needs 5,000 to produce enough tomatoes for his pizza sauce. He wants to divide the remaining 3,000 acres into 5-acre plots, build houses and sell them to Honduran farmers with no money down.

But the main thrust of Domino's farming activities is aimed at benefiting farmers in the United States. This farm and Domino's facilities, which will include a convention center and hotel, will permit us to convene the first state and even national conferences that are truly for small farmers in the history of this country. During my professional career, I have

Tom Monaghan and a speechless Booker Whatley inspect the statue of Whatley.

"Prairie House"—Corporate headquarters of Domino's Pizza. The flags represent the seven countries in which Domino's Pizza is sold.

Whatley explains the finer points of his diversified, high-value farming system to visiting farmers after trading in his straw hat for a Domino's Farms baseball cap.

'THE ULTIMATE SMALL FARM'

"The Whatley Family"—Booker and Lottie are surrounded by Booker's sisters, nieces, grandnephews and other relatives at the Whatley/Domino's farm dedication.

Plaque on the base of Whatley's statue speaks for itself.

Whatley's statue is flanked by three flagpoles, one each for the flags of Domino's Pizza, Whatley Farms and Domino's Farms.

attended many, many so-called "small-farm conferences" and there wasn't a small farmer within 100 miles of the place. That's not going to happen here!

The Domino's farm on Warren Road is dedicated to saving 100,000 small farms by the year 2000. With the resources at hand here, Tom Monaghan's commitment to the idea and the growing number of people raising high-value, pick-your-own crops, we may just wind up with 200,000 of these farms by then. And that would be just fine with me!

Small farms are the very foundation of civilization. We must always remember that and the pivotal role that small farms have played in the rise — and fall — of civilizations as mighty as the Roman Empire. President Harry S. Truman put it so well when he said that people who do not know their history shall forever remain children. That is why I would like to close this final chapter with the following excerpt from the very first report of our first commissioner of agriculture, Isaac Newton, to President Abraham Lincoln in 1862:

"As history is philosophy teaching by example, it would be highly instructive to discuss the condition of influence of agriculture as exhibited in the life of the nations of classical antiquity. . . . After a splendid career of prosperity, filling the world with her fame, Rome culminated and declined. No historical proposition is more susceptible of proof than that the great causes of that decline were the laws enacted affecting real estate and the condition, skill and products of labor. For many years after Rome had grown to greatness, the cultivation of the soil was not only deemed honorable, but was regulated by law, in order that agriculture might yield the largest return to labor and be, in reality, the great conservator of the empire. Not only were flocks and herds kept for food and raiment, and alluvial lands tilled, but the soils in more unfavorable regions were carefully and skillfully cultivated. At first, the allotment of land to each citizen was but 6 acres. It was not ploughed, but spaded, and the yield was very great. Virgil, Cato, and Columella, Rome's chief agricultural writers, invariably urge the cultivation of small farms, in order that the tillage may be thorough. The subdivision of estates, the limitation of their extent, and the habit of personal attention to farming, were excellent conditions for success. "The Romans," says Frederick Von Schlegel, referring to the last days of the republic, "were a thoroughly agricultural people."

Changing this splendid basis of prosperity, permanency, and power, whereby, resting in this soil, Rome pierced the heavens by the force of thought, she grew proud and oppressive, the reins of power slipped from the hands of the middle classes; labor became disreputable; the soil a monopoly, and the masses of the people reckless, unpatriotic, and degraded. A few proprietors held the land and owned the labor. The poverty of the many, with it evils of want, of ignorance, and dependence, existed by the side of the excessive wealth and culture of the few. The lands in Italy and in the conquered provinces were apportioned among the families of the great, instead of being given or sold as free homesteads to the poor. By this unequal distribution of property, and by forcing the husbandman into the army and buying up or taking his land, much of the soil was cultivated by servile labor.

This monopoly of the land and condition of labor operated unfavorably to agriculture, and thus to the prosperity of permanency of the empire. These two causes were destructive to intelligent, interested, and really productive agriculture. Certain staples, it is true, were raised in vast quantities; but these required little skill, and prevented the cultivation of a variety of crops. Old and exhausted lands were abandoned without any attempt to renew their fertility. The laborer felt no moneyed interest, no personal pride, in the result of his toil, and all generous progress in agriculture was retarded.

The Whatleys and a group of about 40 relatives and friends dig into a lunch of (what else?) Domino's Pizza during a tour of Domino's Classic Car Museum in Ann Arbor, Mich., as part of the farm dedication festivities.

'THE ULTIMATE SMALL FARM'

Whatley and Dr. Leon Kneebone, a retired mushroom specialist from Pennsylvania State University, discuss setting up commercial mushroom production facilities on the Whatley/Domino's farm.

The voice of history proclaims, in the clearest manner, that free labor and ownership of the soil by the laborer, if possible, are necessary conditions to the highest success in agriculture and national prosperity. Give the laborer no interest, prospective or otherwise, in the soil he tills, and he cannot be otherwise than wasteful and inefficient. . . .

In the earlier days of the empire, the maximum limitation of freeholds to 500 acres, in connexion with the old Roman love of agriculture, led to a careful and exact mode of agriculture. But in the later days of the empire, says Hallam, "the laboring husbandman, a menial slave of some wealthy senator, had not even the qualified interest in the soil which the tenure of villanage (sic) afforded to the peasant of feudal ages." At this period, notwithstanding Rome's matchless soil and climate, she was compelled to import food from her conquered provinces. . . .

"Owing to the degradation of labor," says Gibbon, "the plebeians disdained to work with their hands, and the husbandman, being obliged to abandon his farm during the term of his military service, soon lost his zest for work. The lands of Italy, which had been originally divided among the families of free and indigent proprietors, were insensibly purchased or usurped by the avarice of the nobles. In the age which preceded the fall of the republic, it was computed that only 2,000 citizens were possessed of any independent subsistence. When the prodigal, thoughtless commons had imprudently alienated not only the use but the *inheritance* of power—to wit, their own homesteads and free life—they sank into a vile and wretched populace!"

Such is one of the great lessons of history; and any nation that desires permanent prosperity and power should learn it well, wisely protecting labor and capital, and encouraging the division and cultivation of the soil. . . .

Agricultural pursuits tend to moderate and tranquilize the false ambition of nations, to heal sectional animosities, and afford a noble arena for honorable rivalry. The acquisition of comparatively slow, but sure, wealth, drawn from and reinvested in the soil, develops health of body, independence and simplicity of life, and love of country; while the rapid accumulation of wealth, not by production, but by trade and speculation, is unnatural and unhealthful. It attracts men to cities and tempts to wild investments. It often unsettles moral principle, and substitutes selfishness for patriotism. Men of the country, living in calm content, and forming almost the entire wealth and population of the Union, constitute the truly conservative element in our politics. The men of the city, living in the midst of excitements, political, social, monetary, and moral, too often feed those baneful causes of national ruin, to wit: speculation, luxury, effeminancy, political corruption, and personal ambition. Never was a truer or more comprehensive line of poetry penned than that which declares that: "God made the country—man made the town."

Next after moral and intellectual forces, home and foreign commerce, manufactures, lines of intercommunication and agriculture form the great arch of our national prosperity—agriculture being the keystone as well as the foundation of all. Agriculture furnishes the food of the nation, the raw materials of commerce and manufactures, and the cargoes of domestic and foreign commerce. It is the cause and evidence of true civilization; for, when tillage begins barbarism ends, and the various arts commence. When agriculture prospers, all other interests prosper. When this fails, depression, panic, ruin, ensue. . . .

Old Rome, with all her elements of decay constantly at work, lasted nearly 1,000 years, and carried her culture, civilization, and arms to a wondrous pitch of glory. May we not hope and devoutly pray that, taking warning from history and the signs of the times, our republic may so learn lessons of wisdom, that, eradicating all destructive tendencies, she will fortify herself against decay, and become what Rome was not—eternal!"

Appendices

I. Fencing Supplies, Information For Intensive Rotational Grazing.

Brookside Industries
Box 158
Tunbridge, VT 05077
(802)889-3737

Charles Kendall
Kencove
111 Kendall Lane
Blairsville, PA 15717
1-800-245-6902
1-800-442-6823 (in Pa.)

Kiwi Fence
RD 2, Box 51A
Waynesburg, PA 15370
(412)627-8158

Premier Fence Systems
RR 1, Box 159
Washington, IA 52353
(319)653-6631 (or 6634)

Snell Systems
P.O. Box 708910
San Antonio, TX 78270
(512)494-5211

Techfence
P.O. Box A-PN
Marlboro, NJ 07746
(201)462-6101

Suggested Reading

Greener Pastures on Your Side of the Fence — Better Farming with Voisin Grazing Management
1986, by Bill Murphy $14.95 (paperback)
Arriba Publishing
213 Middle Rd.
Colchester, VT 05446

Intensive Grazing Management: Forages, Animals, Men, Profits
1987, by Burt Smith, Pingsun Leung & George Love
$29.95, plus $1.50 postage & handling (hardback)
Kingsbery Communications
13235 Stairock
San Antonio, TX 78248

Controlled Grazing Times
Newsletter on intensive, rotational grazing.
For free sample copy, write:
Controlled Grazing Times
13235 Stairock
San Antonio, TX 78248

The Sheep Raiser's Manual
1985, by William Kruesi
$13.95 (paperback)
Williamson Publishing
Box 185, Dept. 8057
Charlotte, VT 05445

II. General Seed Houses
Vegetables and Flower Seeds

Abundant Life Seed Foundation
P.O. Box 772
Port Townsend, WA 98368
(206) 385-5660
Organic and untreated seeds: vegetables, herbs, wildflowers, trees. All open-pollinated varieties. $5 for one-year subscription, $6 for Canadian subscribers. Subscriptions includes combined seed and book catalog. Sample catalog $1.

Burgess Seed & Plant Co.
905 Four Seasons Road
Bloomington, IL 61701
(309) 663-9551
Vegetable seeds, dwarf and standard fruit trees, nut trees, berries. Free catalog.

W. Atlee Burpee Co.
300 Park Ave.
Warminister, PA 18974
Flowers, vegetables, fruits, nursery stock. Free catalog.

D. V. Burrell Seed Growers Co.
P.O. Box 150
Rocky Ford, CO 81067
(303) 254-3318
Vegetable and flower seeds. Free catalog.

De Giorgi Co. Inc.
P.O. Box 413, 1411 Third St.
Council Bluffs, IA 51502
(712) 323-2372
Vegetable and flower seeds. Catalog $1.

Farmer Seed & Nursery
818 N.W. Fourth St.
Faribault, MN 55021
(507) 334-1623
Vegetable and flower seeds, fruit trees and small fruits, shade trees, shrubs, flower bulbs, roses, perennials. Free catalog.

Henry Field Seed & Nursery Co.
2176 Oak St.
Shenandoah, IA 51602
(712) 246-2110
Vegetables, fruits, trees, shrubs, perennials. Spring and fall bulbs, herb seeds and plants. Free catalog.

Gurney Seed & Nursery Co.
Second and Capitol
Yankton, SD 57079
(605) 665-4451
Complete offering—flower and vegetable seeds, nursery stock. Free catalog.

Harris Seeds
Moreton Farm
3670 Buffalo Road
Rochester, NY 14624
(716) 594-9411
Vegetable and flower seeds. Free catalog.

Jung Seed Co.
335 S. High St.
Randolph, WI 53957
(414) 326-3121
Garden, farm and flower seeds, ornamental shrubs and plants. Free catalog.

Orol Ledden & Sons
P.O. Box 7
Sewell, NJ 08080-0007
(609) 468-1000
Vegetable seeds and plants, farm seeds, annual and perennial flower seeds. Free catalog.

Letherman's Inc.
1221 Tuscarawas St. E.
Canton, OH 44707
(216) 452-5704; (216) 452-8866
Vegetable and flower seeds, grass seed, turf products. Free catalog.

Earl May Seed & Nursery Co.
208 N. Elm St.
Shenandoah, IA 51603
(712) 246-1020
Complete offering of flower, vegetable and lawn seeds, vegetable plants, bulbs, perennials, trees, shrubs, evergreens, roses. Free catalog.

Mellinger's Inc.
2310 W. South Range Road
North Lima, OH 44452-9731
(216) 549-9861
Full line of country-living items from seed, plants, trees and bushes to fertilizers, tools, kitchen utensils, books. More than 4,000 items in free catalog. Or rushed by first class mail, $1.50.

Necessary Trading Co.
New Castle, VA 24127
(703) 864-5103
Full line of green manure and cover crop seeds, specialty fruit and vegetable equipment. Emphasis on biological soil fertility management and natural pest control. Catalog $2, deductible from first order.

Park Seed Co. Inc.
Highway 254 North
Greenwood, SC 29647-0001
(803) 223-7333
Flower and vegetable seeds, bulbs, perennial plants and houseplants, herbs, wildflowers. Free catalog.

Pony Creek Nursery
Tilleda, WI 54978
(715) 787-3889
Vegetable and flower seeds, trees, shrubs. Free catalog.

Porter & Son, Seedsmen
1510 E. Washington St.
P.O. Box 104
Stephenville, TX 76401-0104
Vegetable and flower seeds. Specializes in Texas and Southwestern varieties. Free catalog.

Seedway Inc.
Box 250
Hall, NY 14463
(716) 526-6391
Vegetable and flower seeds. Free catalog.

R. H. Shumway, Seedsman
P.O. Box 777
Rockford, IL 61105
and P.O. Box 1
Graniteville, SC 29829
(803) 663-6276
Vegetable, flower, farm, lawn and field seeds. Fruit trees, berries, roses bulbs, ornamental shrubs. Specialize in old-time and open-pollinated seeds. Wholesale and retail. Catalog $1, deductible from first order, includes packet of exclusive ABE LINCOLN tomato seed.

Stokes Seeds Inc.
Box 548
Buffalo, NY 14240
and 28 Water St.
Fredonia, NY 10463

Stokes Seeds Ltd.
39 James St., Box 10
St. Catherines, Ontario
Canada L2R 6R6
(416) 688-4300
Vegetable and flower seeds. Free catalog.

T & T Seeds Ltd.
111 Lombard Ave., Box 1710
Winnipeg, Manitoba
Canada R3C 3P6
(204) 943-8483
Vegetable, flower and grass seed, perennials, bulbs and rose bushes, shrubs, fruit, shade trees. Catalog $1.

Thompson & Morgan Inc.
Farraday and Gramme Avenues
P.O. Box 1308
Jackson, NJ 08527
(201) 363-2225
Vegetable and flower seeds. Free catalog.

Otis S. Twilley Seed Co.
P.O. Box 65
Trevose, PA 19047
(215) 639-8800
Vegetables, flower and herb seeds, bulbs, grasses and farm seeds. Free catalog.

Wyatt-Quarles Seed Co.
P.O. Box 739
Garner, NC 27529
(919) 832-0551
Vegetable, flower and herb seeds, bulbs, grasses and farm seeds. Free spring seed catalog and fall bulb catalog.

III. Vegetable Specialties
Unusual, Regional, Ethnic

Walter Baxter Seed Co.
P.O. Box 8175
Weslaco, TX 78596
(512) 968-3187
Will fill orders, both large and small, for most any vegetable grown in Texas and surrounding areas. Price list available.

Becker's Seed Potatoes
RR #1
Trout Creek, Ontario
Canada P0H 2L0
(705) 724-2305
Hard-to-obtain licensed Canadian potato varieties, old and new, including YUKON GOLD, GREEN MOUNTAIN, CARIBE and RED GOLD. Free price list.

Brown's Omaha Plant Farms Inc.
P.O. Box 787
Omaha, TX 75571
(214) 884-2421
Onion plants. Price list 25 cents.

Butterbrooke Farm Seed Co-op
78 Barry Road
Oxford, CT 06483
(203) 888-2000
Specializes in chemically untreated, pure line, short-maturity vegetable seeds at reduced prices. Same-day service. Free seed list.

The Cook's Garden
Box 65
Londonderry, VT 05148
(802) 824-3400
Salad lovers' catalog. Gourment greens and more than 30 varieties of lettuce. Large selection of Italian vegetables. Catalog $1.

Epicure Seeds Ltd.
P.O. Box 450
Brewster, NY 10509
(800) 833-2900, toll-free
(914) 279-4901, in New York
Imported and unusual vegetable seeds. Free catalog includes recipes for several specialty vegetables.

Far West Fungi
P.O. Box 428
So. San Francisco, CA 94080
Complete selection of mushroom-growing kits and mushroom spawn — shiitake, button, almond, tree-oyster mushrooms and more. For catalog send self-addressed, stamped business envelope.

Fisher's Garden Store
P.O. Box 236
Belgrade, MT 59714
(406) 388-6052
Vegetable and flower seeds for high altitudes or short growing seasons. Supplies Northwest only. Free catalog.

Fred's Plant Farm
Route 1, P.O. Box 707
Dresden, TN 38225
(901) 364-5419
Sweet potato seeds and plants; tobacco seeds and products. Send Self-addressed, stamped business-size envelope for the price list.

Garden City Seeds
Box 297
Victor, MT 59875
(406) 961-4837
Part of a non-profit organization dedicated to self-reliance. Specializes in older, short-season varieties. Emphasis placed on organic growing and open-pollinated varieties suitable for the North. Catalog $1.

Golden Acres Farm
RFD 1, Box 1462
Western Ave.
Fairfield, ME 04937
(207) 453-7771
Vegetable and herb varieties selected for their adaptation to the Northeast. Free catalog.

Good Seed Co.
P.O. Box 702
Tonasket, WA 98855
Untreated, open-pollinated seeds for the North, featuring heirloom vegetables, native American maize, tepary beans, herbs, flowers, garlic. Catalog $2. Price list 25 cents. Trade/barter welcome.

H. G. Hastings Co.
P.O. Box 4274
Atlanta, GA 30302-4274
(404) 524-8861
Specializes in vegetable seeds and nursery stock for the South; flowers, fruit and nut trees. Free catalog.

Herb Gathering Inc.
Tastefully Creative Seeds
5742 Kenwood Ave.
Kansas City, MO 64110
(816) 523-2653
French vegetable and herb seeds. Special "Creation Garden" section offers seed for craftspeople. Herb books. Unique "fresh herbs to your door" service. (Formerly Demonchaux Seeds.) Catalog $2, deductible from first order.

High Altitude Gardens
P.O. Box 4238
Ketchum, ID 83340
(208) 726-3221
Varieties tested and selected for high altitude or cold climates. Vegetables, wildflowers, native grasses and herbs. Some vegetable varieties available in commercial quantities. Books, tools, pest controls. Catalog $2.

Horticultural Enterprises
P.O. Box 810082
Dallas, TX 75381-0082
More than 30 varieties of peppers, both hot and sweet. Also jicama and several herbs used in Mexican cooking. Free price list.

Ed Hume Seeds Inc.
P.O. Box 1450
Kent, WA 98032
(206) 859-1110
Vegetable and flower seeds for cool, short-season climates, featuring WALLA WALLA onion, SUGAR DADDY BUSH stringless snap pea, SUPERSNOOP bush sweet peas. Recommendations for Alaskan growers. Free catalog.

Illinois Foundation Seeds
Box 722
Champaign, IL 61820
(217) 485-6260
X-TRA SWEET sweet corn seeds. Free price list.

Le Jardin Du Gourmet
P.O. Box 30
West Danville, VT 05873
Shallots, leeks, garlic, herbs, French vegetable seeds. Catalog plus eight sample packs of herb seeds — $1.

Johnny's Selected Seeds
310 Foss Hill Road
Albion, ME 04910
(207) 437-9294 — Customer Service
(207) 437-4301 — Orders
Vegetable, herb, flower and specialty seeds for cool-climate growers. Free catalog features seed-starting supplies, biological and botanical insecticides, watering equipment, season-extending supplies and a wealth of cultural and varietal information. One of the best, most technically complete and helpful catalogs in print.

Kalmia Farm
P.O. Box 3881
Charlottesville, VA 22903
Hard-to-find shallots, garlic varieties and multiplier onions, including the potato and white multiplier onions. Saffron crocus and flowering alliums. Also, books on the history and growing of onions and garlic. Free catalog.

Kilgore Seed Co.
1400 W. First St.
Sanford, FL 32771
Specializes in vegetable and flower seeds for Gulf Coast states and the sub-tropics, but ships to 50 states. Catalog $1, deductible from first order.

Kitazawa Seed Co.
1748 Laine Ave.
Santa Clara, CA 95051
(408) 249-6778
Japanese and Chinese vegetable seeds. Price list.

III. VEGETABLE SPECIALTIES

Lagomarsino Seeds
5675-A Power Inn Road
Sacramento, CA 95824
(916) 381-1024
A pioneer seed company in business since the 1890's. Specializes in older varieties and European vegetables. Also carries flower seeds, pasture seed and cover crops with a full line of rhizobial inoculants. Free catalog.

Liberty Seed Co.
Box 806
New Philadelphia, OH 44663
(216) 364-1611
Specializes in vegetable, flower and herb seeds for the serious grower. Commercial flower and bedding catalog, commercial vegetable catalog, home garden catalog; free.

Le Marche Seeds International
P.O. Box 190
Dixon, CA 95620
(916) 678-9244
Specialty vegetable and herb seeds from around the world. Wide selection of lettuces, peppers and Italian chicories. Good range of old-fashioned, open-pollinated varieties plus new, productive hybrids. Catalog includes botanical and cultural details as well as interesting recipes; $2.

Margrave Plant Co.
117 Church St.
Gleason, TN 38229
(901) 648-5174
Grower and shipper of sweet potato plants since 1933. Free price list.

Moutain Valley Seeds and Nursery
1798 North 1200 East
North Logan, UT 84321
(801) 752-0247
Varieties suited to Utah and Intermountain West. Plant protection devices. Free catalog.

Mushroom People
P.O. Box 159A
Inverness, CA 94937
(415) 663-8504
A clearinghouse of mushroom information. Specializes in shiitake spawn; consulting service available. Complete catalog includes cultures, supplies and books; $2.

Nichols Garden Nursery
1190 North Pacific Hwy.
Albany, OR 97321
(503) 928-9280
New and traditional vegetables for the flavor-conscious gardener. Gourmet and Oriental varieties, herb plants, herb and flower seeds. Free catalog.

The Pepper Gal
10536 119th Ave. N.
Largo, FL 33543
200 pepper varieties — hot, sweet and ornamental. Free price list.

Piedmont Plant Co. Inc.
Box 424-D
Albany, GA 31703
(912) 883-7029
Specializes in field-grown tomato, pepper, onion, cabbage and eggplant plants for spring planting. Also sells broccoli and cauliflower plants. Free catalog.

Pinetree Garden Seeds
New Gloucester, ME 04260
(207) 926-3400
Specializes in small packets of vegetable, herb, flower and houseplant seeds at reduced prices. Space-saving varieties and a large selection of gardening books. Free catalog.

Redwood City Seed Co.
P.O. Box 361
Redwood City, CA 94064
(415) 325-7333
Offers vegetables and herbs from around the world. Broccoli, cabbages, cucumbers, greens, daikon radishes and spinach from the Orient; beans, corns, hot and sweet peppers from North and Central America; lettuces, melons and herbs from Europe. Catalog $1.

Shepherd's Garden Seeds
7389 West Zayante Road
Felton, CA 95018
(408) 335-5400
Specializes in European vegetable seeds. Offers individual varieties or special collections of French, Italian, Oriental and Mexican vegetables. Also herb and salad garden collections, edible flowers, baby vegetables, everlasting and "cottage garden" flowers. Catalog includes growing hints, companion-planting tips and recipes; $1.50.

Siberia Seeds
Box 2528
Olds, Alberta
Canada T0M 1P0
(403) 556-7333
Heirloom tomato seed varieties. Supplier of SIBERIA and other tomato varieties suitable for Northern climates. Send a self-addressed, stamped business-size envelope.

Southern Exposure Seed Exchange
P.O. Box 158
North Garden, VA 22959
Specializes in open-pollinated, disease-resistant vegetable varieties. All seed is untreated and much of it is grown organically. Heirloom seeds, multiplier onions, herbs, sunflowers and uncommon vegetables. New for 1987 — heirloom apple trees. Catalog and growing guide $2.

Steele Plant Farm
Box 987
Gleason, TN 38229
(901) 648-5476
Vegetable plants. Specializes in sweet potatoes — new quick-maturing "vineless" varieties, prolific in Northern gardens. Mixed dozens (four plants each of three varieties). Specific growing guide for each state. Send two first-class stamps for catalog.

Sunrise Oriental Seed Co.
P.O. Box 10058
Elmwood, CT 06110-0058
(203) 666-8071
More than 140 varieties of Oriental vegetables. Cooking and gardening books. Sprouting seed and unusual Oriental plants. Catalog $1, deductible from first order.

Territorial Seed Co.
P.O. Box 27
Lorane, OR 97451
(503) 942-9547
Primarily for growers west of the Cascades, and on the lower mainland and islands of British Columbia. Unique American and European seed selected for short seasons, cool climates. Many overwintered varieties. Free catalog.

Tomato Growers Supply Co.
P.O. Box 2237
Fort Myers, FL 33902
126 tomato varieties. Cooking equipment, books about tomatoes. Free catalog includes information about growing and selecting tomato varieties.

Tsang & Ma International
P.O. Box 294
Belmont, CA 94002
(415) 595-2270
Oriental vegetable seeds. Oriental cooking tools, sauces, oils and seasonings. Free brochure.

Vermont Bean Seed Co.
Garden Lane
Fair Haven, VT 05743-0250
(802) 265-4212
Complete seed line featuring the world's largest selection of beans, but don't let the name fool you. VBS has just about any other kind of seed you could want, too. Free catalog.

Willhite Seed Co.
P.O. Box 23
Poolville, TX 76076
(817) 599-8656
Superior-quality vegetable seeds, more than 300 varieties. Specializes in watermelon and cantaloupe varieties. Free catalog.

IV. Herb Seeds and Plants

ABC Nursery & Greenhouse
Route 1, Box 313
Lecoma, MO 65540
(314) 435-6389
Herb plants, herb collections. Send first-class stamp for price list.

Capriland's Herb Farm
534 Silver St.
Coventry, CT 06238
(203) 742-7244
Herb plants, seeds, wreaths, potpourris and other herb products. Garden ornaments and books. Lecture-luncheon programs conducted at farm, April-December. Reservations required. Open 9-5 daily year-round. Free brochure.

Casa Yerba Gardens
Star Route 2, Box 21
Days Creek, OR 97429
(503) 825-3534
Rare and unusual herb seeds and plants, grown organically. Catalog $1.

Catnip Acres Herb Farm
67-R Christian Street
Oxford, CT 06483
(203) 888-5649
More than 200 varieties of herb seeds; many not available elsewhere. Hundreds of varieties of herb plants, including more than 100 types of scented geraniums available at the nursery only. Wreaths, potpourris, dried herbs, books and more. Lectures and wreath-making workshops. Open Tuesday—Sunday 10-5, April-Christmas. Seed catalog $1; price list free.

Companion Plants
Route 6, Box 88
Athens, OH 45701
(614) 592-4643
More than 400 varieties of herb plants and seeds. Specializes in exotic herbs and useful and native plants. Catalog $2, deductible from first order. Wholesale listing available.

Earthstar Herb Gardens
438 W. Perkinsville Road
Chino Valley, AZ 86323
(602) 636-2565
Herb plants, products, seeds and dried herbs. Scented geraniums and dried flowers. Catalog $1, deductible from first order.

Fox Hill Farm
444 W. Michigan Ave., Box 7
Dept. TNF
Parma, MI 49269-0007
(517) 531-3179
More than 350 varieties of quality herb plants: culinary, medicinal, dye, everlasting and insect-repelling; scented geraniums. Fresh herb produce. Condiments. Printed material. Catalog $1; free Herb Shopping List.

Goodwin Creek Gardens
Box 83
Williams, OR 97544
Organic growers of more than 200 varieties of herbs, both plants and seeds. Everlastings. Catalog 50 cents.

Hartman's Herb Farm
Old Dana Road
Barre, MA 01005
(617) 355-2015
More than 300 varieties of herbs and herb products. Live plants, dried arrangements, wreaths, potpourri, oils. Display garden for visitors. Catalog $1.

Hemlock Hill Herb Farm
Hemlock Hill Road
Litchfield, CT 06759-0415
(203) 567-5031
Perennial and biennial herb plants. Catalog 50 cents.

Meadowbrook Herb Garden
Route 138
Wyoming, RI 02898
(401) 539-7603
Herb and wildflower seeds, organic fertilizers and pest controls, organically grown herb seasonings and teas, herb jellies and vinegars, herbal skin care products, books. Catalog $2.

Sandy Mush Herb Nursery
Route 2, Surrett Cove Road
Leicester, NC 28748
(704) 683-2014
Perennial herb plants, scented geraniums — over 600 varieties — along with books, seeds, potpourri, insect repellents, growing and selection guides. Catalog/handbook $2, deductible from first order.

Sunnybrook Farms Nursery
9448 Mayfield Road
Chesterland, OH 44026
(216) 729-7232
Herb plants and seeds, scented geraniums, ivies, herbal products. Catalog $1, deductible from first order.

Taylor's Herb Gardens Inc.
1535 Lone Oak Road
Vista, CA 92083
(619) 727-3485
130 varieties of herb plants, 180 varieties of herb seeds. Catalog $1.

Well-Sweep Herb Farm
317 Mt. Bethel Road
Port Murray, NJ 07865
(201) 852-5390
Herb plants and seeds, scented geraniums, perennials, dried flowers. Catalog $1.

V. Berry and Small-Fruit Specialties

A.G. Ammon Nursery Inc.
Route 532, Box 488
Chatsworth, NJ 08019
(609) 726-1370
Certified blueberry plants. Free price list.

Ahrens Strawberry Nursery Inc.
Route 1
Huntingburg, IN 47542
(812) 683-3055
Strawberry, raspberry, gooseberry, blueberry and blackberry plants. Currants, grapes and rhubarb. Fruit trees. Free catalog.

Allen Co.
Box 1577
Salisbury, MD 21801-1577
(301) 742-7122
Specializes in strawberry plants; also blueberries, red raspberries, thornless blackberries, asparagus. Features ALLSTAR strawberry. Free catalog.

Brittingham Plant Farms
P.O. Box 2538
Salisbury, MD 21801
(301) 749-5153
Strawberry, blueberry, raspberry, thornless blackberry and grape plants. Asparagus roots. Free catalog.

Edible Landscaping
Route 2, Box 343A
Afton, VA 22920
(703) 949-8408
Hardy kiwis, hardy Oriental persimmons, black currants, gooseberries, mulberries, jujube, apples, figs, grapes, thornless blackberries. Free price list.

Enoch's Berry Farm
Route 2, Box 227
Fouke, AR 71837
(501) 653-2806
SHAWNEE (Pat. #3858), CHEYENNE and ROSBOROUGH blackberry plants and root cuttings. Free price list.

Finch Blueberry Nursery
P.O. Box 699
Bailey, NC 27807
(919) 235-4664
Rabbiteye blueberry plants. Free brochure and price list.

Dean Foster Nurseries Inc.
P.O. Box 127
Hartford, MI 49057
(616) 621-2419
Specializes in strawberries. Also raspberries, blackberries, currants, gooseberries, asparagus, rhubarb, fruit trees. Free catalog.

Hartmann's Plantation Inc.
P.O. Box E
Earleton, FL 32631
(904) 468-2087
Blueberry plants, extensive listing. Free catalog.

House of Wesley, Nursery Division
2200 E. Oakland Avenue
Bloomington, IL 61701
(309) 663-9551
General line of nursery stock, strawberries, grapes, dwarf fruit and nut trees. Free catalog.

Ison's Nursery & Vineyard
Brooks, GA 30205
(404) 599-6970
Specializes in muscadine grapes, blackberries, blueberries, raspberries, strawberries, fruit and nut trees. Free catalog.

Wm. Krohne Plant Farms
Strawberry Plant Nursery
Route 6, Box 586
Dowagiac, MI 49047
(616) 424-3450
Features 17 varieties of strawberry plants. Free brochure.

Nourse Farms Inc.
Box 485, RFD
South Deerfield, MA 01373
(413) 665-2658
Strawberry and raspberry plants, asparagus and rhubarb roots. Free catalog.

Lon's Oregon Grapes
4285 Portland Road N.E.
Salem, OR 97303
(503) 393-5165
More than 50 grape varieties available. Testing and consultation. Free catalog. Send self-addressed, stamped business-size envelope.

Makielski Berry Farm & Nursery
7130 Platt Road
Ypsilanti, MI 48197
(313) 434-3673
Fruit trees and small fruits; specializes in raspberries. TRISTAR day-neutral strawberry. Also currants, gooseberries, thornless blackberries, seedless grape varieties and asparagus roots. Free catalog.

Rayner Bros. Inc.
P.O. Box 1617
Salisbury, MD 21801
(301) 742-1594
Specializes in strawberry, blueberry and asparagus plants; also raspberries, bush fruits, fruits, drawf fruit trees, evergreen seedlings. Free catalog.

Stanley & Sons Nursery
11740 S.E. Orient Drive
Boring, OR 97009
(503) 663-4391
50 varieties and species of hardy kiwis; ornamental nursery stock, conifers and Japanese maples. Free catalog — specify kiwi or conifers.

Square Root Nursery
4764 Deuel Road
Canandaigua, NY 14424
(716) 394-3140
34 grape varieties, propagated and grown on premises. Every order includes guide "How to Grow Grapes." Free catalog.

VI. Fruit and Nut Tree Nurseries

Adams County Nursery Inc.
P.O. Box 108
Aspers, PA 17304
(717) 677-8105
Fruit trees. Free catalog.

Amberg's Nursery Inc.
3164 Whitney Road
Stanley, NY 14561
(716) 526-5405
Growers of size-controlled fruit trees. Good selection of horticultural tools. Free price list.

Bear Creek Nursery
P.O. Box 411
Northport, WA 99157
Large selection of antique apple varieties. Rare and valuable nut tree varieties. Windbreak, wildlife, conservation and rootstock plants, organically grown. Specializes in hardy and drought-tolerant plants. Send two first-class stamps for catalog.

Burnt Ridge Nursery
432 Burnt Ridge Road
Onalaska, WA 98570
(206) 985-2873
Specializes in grafted nut trees from select Northern cultivars. Also offers hardy kiwis, loquats, pineapple guavas, mulberries and figs. Send self-addressed, stamped business-size envelope for price list.

Chestnut Hill Nursery Inc.
Route 1, Box 341
Alachua, FL 32615
(904) 462-2820
Hybrid American Chinese chestnuts and Oriental persimmons. Free brochure and price list.

Cumberland Valley Nurseries
Box 471
McMinnville, TN 37110
(615) 668-4153
Specializes in peach, plum or nectarine trees on LOVELL-HALFORD or NEMAGUARD rootstocks. Limited quantity of pecan, apple, pear and cherry trees. Free price list.

Fig Tree Nursery
P.O. Box 124
Gulf Hammock, FL 32639
(904) 486-2930
12 varieties of figs, muscadine grapes, poppy seed, pepper seed, mulberry trees. Price list available for one first-class stamp. Does not ship to California or Arizona.

Friends of the Trees Seed Service
P.O. Box 1466
Chelan, WA 98816
Supplies tree, shrub and herbaceous perennial seeds. Also manages a perennial seed exchange. *Friends of the Trees Yearbook* is both a seed catalog and an information directory; includes Perennial Seed Exchange lists, lists of regional organizations, seed and plant exchanges, searches and mail-order companies, much more. 1986 Yearbook $4.

Fowler Nurseries Inc.
525 Fowler Road
Newcastle, CA 95658
(916) 645-8191
Fruit and nut trees, berries, grapes. Catalog $2.

Louis Gerardi Nursery
1700 E. Highway 50
O'Fallon, IL 62269
(618) 632-4456
Offers scion wood for varieties of mulberries, pecans, hickories, Persian walnuts, black walnuts, heartnuts, hicans, butternuts, Chinese chestnuts, filberts and American persimmons. For free catalog send a self-addressed, stamped business-size envelope.

Hilltop Trees
The Nursery Corp.
P.O. Box 578
Hartford, MI 49057
(616) 621-3135
Dwarf and standard fruit trees. Catalog's handbook section includes information on pollination, pruning and choosing rootstocks. Minimum order 10 trees. Free catalog.

Johnson Nursery Inc.
#5, Box 29J
Ellijay, GA 30540
(404) 273-3187
Dwarf, semi-dwarf and standard fruit trees — apple, peach, pear; specializes in antique peaches; also carries grapes, blueberries. Free catalog.

Lawson's Nursery
Route 1, Box 294, Yellow Creek Road
Ball Ground, GA 30107
(404) 893-2141
Specializes in old-fashioned and unusual fruit trees. Good selection of antique apple varieties grafted on (Oregon certified virus-free) superdwarf rootstock. Grapes and blueberries. Free catalog.

Henry Leuthardt Nurseries
Montauk Highway, P.O. Box 666-0
East Moriches, NY 11940
(516) 878-1387
Dwarf and semi-dwarf fruit trees, espaliered fruit trees, hybrid grapes, berry plants. Free price list; handbook $1, deductible from first order.

Living Tree Centre
P.O. Box 797
Bolinas, CA 94924
(415) 868-2224
More than 65 historic apple, pear, apricot and persimmon varieties. Features REINETTE SIMIRENKO apple and WARREN pear. *The Living Tree Journal* also contains growing information and articles; Journal/catalog $6, deductible from purchase of a tree.

J. E. Miller Nurseries Inc.
5060 W. Lake Road
Canadaigua, NY 14424
(800) 462-9601 in N.Y.; (800) 828-9630
Fruit and nut trees, grape vines, small fruits, berry bushes, ornamentals. Free catalog.

Musser Forests Inc.
P.O. Box 15 MA
Indiana, PA 15701-0340
(412) 465-5685 or 465-5686
Evergreen and hardwood seedlings and transplants. Christmas tree seedlings. Landscaping shrubs, ground covers, windbreaks. Free catalog.

Nolin River Nut Tree Nursery
797 Port Wooden Road
Upton, KY 42784
(502) 369-8551
More than 100 varieties. Budded or grafted nut trees. Black and Persian walnuts, pecans, butternuts, chestnuts, hickories, heartnuts and hicans. Free price list.

Northwoods Nursery
28696 S. Cramer
Molalla, OR 97038
(503) 651-3737 (Tues.-Sun.)
Unusual, hard-to-find varieties of fruits and nuts; Asian pears, hardy kiwis, grafted native persimmons, figs, woodlot trees. Apprenticeship program. Free catalog.

N.Y. State Fruit Testing Cooperative Association Inc.
P.O. Box 462
Geneva, NY 14456
(315) 787-2205
Apples, crab apples, pears, cherries, plums, peaches, nectarines, apricots, grapes, blueberries, red, black & purple raspberries, blackberries, gooseberries, currants, elderberries, strawberries, rootstocks, propagating wood, bench grafts. Free catalog.

Patrick's Nursery
P.O. Box 130-0
Ty Ty, GA 31795-0130
(912) 382-1770 or 382-1122
Specializes in fruit, nuts, grapes and berry plants for the South. Free catalog.

Peaceful Valley Farm Supply
11173 Peaceful Valley Road
Nevada City, CA 95959
(916) 265-3276
More than 450 varieties of plants, including reforestation and edible landscaping trees, rootstocks, shrubs, tropical and subtropical fruits, small fruits, container-grown citrus, flower bulbs and more. Catalog includes complete line of nursery stock, natural fertilizers, biological controls, hand and power tools, propagation supplies, seeds, farm equipment and growing advice; $2, deductible from first order.

VI. FRUIT AND NUT TREE NURSERIES

Raintree Nursery
391 Butts Road
Morton, WA 98356
(206) 496-5410
Specializes in disease-resistant fruit and nut varieties for organic growers, and a wide variety of edible landscape plants. Free catalog.

St. Lawrence Nurseries
RD#2
Potsdam, NY 13676
(315) 265-6739
Organically grown fruit and nut trees for Northern climates. More than 100 apple cultivars, many heirloom varieties. Cold-hardy black walnuts for quantity plantings. Free catalog.

Southmeadow Fruit Gardens or Grootendorst Nursery
Lakeside, MI 49116
(616) 469-2865
Choice and unusual fruit varieties for the connoisseur and home gardener. Includes 198 apple varieties, rare grapes and gooseberries, pears, medlars, peaches, nectarines, plums, apricots, cherries, quince, currants, conservation fruits and rootstock. Illustrated catalog $8. Free price list.

Stark Bro's Nurseries & Orchards Co.
Box 2281
Louisiana, MO 63353-0010
(314) 754-5511
Specializes in dwarf, semi-dwarf and standard fruit trees, including dozens of exclusive varieties. Also grapes, berries, cherries, nut, shade and ornamental trees, shrubs, roses and more. Free catalog.

Turkey Hollow Nursery
HCR 3, Box 860
Cumberland, KY 40823
(606) 589-5378
Supplies full dwarf and interstem rootstock of more than 200 apple varieties. Limited number of drawf and semi-dwarf trees. Custom propagation. Free price list.

Van Well Nursery Inc.
P.O. Box 1339
Wenatchee, WA 98801
(509) 663-8189
Quality fruit trees: apple, pear, peach, cherry, prune, plum, nectarine, apricot, dwarf trees, berry plants, grapes and nut trees. Free catalog includes both retail and commercial prices.

Windy Hills Farm
1565 E. Wilson Road
Scottville, MI 49454
(616) 757-2373
Specializes in hardwood trees and hardy edible-landscape plantings: organically grown native fruits, nut-tree seedlings, grafted nut trees, honey plants, wildlife plants. Free price list includes species-selection guide.

Womack's Nursery Co.
Route 1, Box 80
De Leon, TX 76444-9660
(817) 893-6497
Large selection of pecan trees. Fruit trees, including peaches, apricots, figs, pears, plums and persimmons. Pruning and propagation tools. Free catalog.

M. Worley Nursery
98 Braggtown Road
York Springs, PA 17372
(717) 528-4519
Dwarf, semi-dwarf and standard fruit trees. Specializes in peach trees and old-time apple varieties. Free catalog.

VII. People Who Can Help

AgMarket Search
RPM Systems Inc.
Suite 200
938 Chapel Street
New Haven, CT 06510
(203) 776-2358

Agroecology Program
Department of Environmental Studies
University of California
Santa Cruz, CA 95064
(408) 429-2634

Alternative Farming System Information Center
National Agriculture Library
Room 111
Beltsville, MD 20705
(301) 344-3755

Alternatives for the 80's Office
628 Clark
University of Missouri
Columbia, MO 65211
(314) 882-2121

Always Buy Colorado (ABC)
Tom Jenkinson
Box 38998
Denver, CO 80238
(303) 451-7721

California Agrarian Action Project
P.O. Box 464
Davis, CA 95617
(916) 756-8518

California Certified Organic Farmers
State Office
P.O. Box 8136
Santa Cruz, CA 95061-8136
(408) 423-2263

California Department of Food and Agriculture
Direct Marketing Program
1220 N. Street
Room 427
Sacramento, CA 95814
(916) 445-9280

California Institute for Rural Studies
P.O. Box 530
Davis, CA 95616
(916) 756-6555

Center for Rural Affairs
Don Ralston
P.O. Box 405
Walthill, NE 68067
(402) 846-5428

Clark's Farm
Richard Clark
111 Worthley Road
Bedford, NH 03102
(603) 623-9567

Coleman's Natural Beef
Mel Coleman
5125 Race Court Unit 4
Denver, CO 80216
(303) 297-9393

VII. PEOPLE WHO CAN HELP

Colorado Agriculture Leadership Program
Louise R. Singleton
1525 Sherman Street
Room 406
Denver, CO 80203
(303) 866-3219

Ecology Action/Common Ground
2225 El Camino Real
Palo Alto, CA 94306
(415) 328-6752

Farmer Directory Marketing Association
Carolyn Goldall, President
Helvetia Good Fruit Company
Route 1
P.O. Box 606
Hillsboro, OR 97124
(503) 647-5323

Farmer Directory Marketing Association
Nancy Harding, Vice President
Overlake Blueberry Farm
1210 170 Avenue N.E.
Bellevue, WA 98008
(206) 641-7685 home

Gasconade Farm
Paydown Road
Vienna, MO 65582
(314) 744-5366

Institute for Alternative Agriculture
Dr. Garth Youngberg
9200 Edmonston Road
Suite 117
Greenbelt, MD 20770
(301) 441-8777

Kirkwood Community College
Curtis Stutzman
P.O. 2068
Cedar Rapids, IA 52406
(319) 398-5609

Maine Organic Farmers and Gardners Associaton
Box 2176
Augusta, ME 04330
(207) 622-3118

Malachite Small Farm School
Box 21
Gardner, CO 81040
(303) 746-2389

New Farm Magazine
222 Main Street
Emmaus, PA 18098
(215) 967-5171

Pennsylvania Farmworker Opportunities (PAFO)
State Operations Office
107 North Front Street
Harrisburg, PA 17101
(717) 232-1931

Rocky Mountain Institute
Dr. Tom McKinney
Drawer 248
Old Snowmass, CO 81654
(303) 927-3851

Rodale Institute/Regenerative Agriculture Association
222 Main Street
Emmaus, PA 18098
(215) 967-5171

Rural Enterprise Magazine
P.O. Box 878
Menomonee Falls, WI 53051
(414) 255-0100

Rural Ventures Inc.
Don Cliff
511 South 11th Avenue
Suite 401
Minneapolis, MN 55415
(612) 375-8050

Howard W. Kerr Jr., Program Director
Office for Small Scale Agriculture
OGPS-USDA
14th and Independence Ave. S.W.
635 Hamilton Bldg.
Washington, D.C. 20250
(202) 535-0234

Small Farm Resources Project
Center for Rural Affairs
P.O. Box 736
Hartington, NE 68739
(402) 254-6893

Small Farm Technical Newsletter
Whately Farms Inc.
7 South Haardt Drive
Montgomery, AL 36105-0827
(205) 265-7786

Small Farms Action Group
Box 80729
Lincoln, NE 68501
(402) 475-3637

Small Farms Center
Christy Wyman
University of California
Davis, CA 95616
(916) 752-0678

Wisconsin Rural Development Center
P.O. Box 504
Black Earth, WI 53515-0504
(608) 767-2539

Index

Advertising, 5, 7, 18, 20-24, 91, 92, 102, 107, 116, 117, 123, 124. *See also* Brochure; Media Day; Newspaper; Poster; Publicity; Radio; Television.
Alfalfa, 9, 58-61, 63, 133
Allelopathy, 132-134. *See also* Weed Control.
Answering Machine, 18, 123. *See also* Telephone.
Aphid, 97, 98, 142, 143
Apiary, 54, 55. *See also* Bee.
Apple, 16, 18, 50, 51, 94-96, 112, 116, 117, 131, 132, 143
Asparagus, 16
Auction, 44-47, 123
Avocado, 50, 72

Basil, 79
Bass, 26, 27. *See also* Fish; Fishing Rights; Pond.
Bean, 4, 34, 42
    Bell, 70, 71, 143
    Lima, 16, 26, 104, 121
    Snap, 16, 132, 133
    Southern, 15
    String, 121, 122
Bedding Plant, 27
Bedmaker, 37, 39, 44
Bee, 3, 5, 55, 57, 74, 88, 89, 98, 131, 151, 154
    Hive, 54, 55, 74, 117
    Honey from, 4, 9, 14, 15, 23, 25-28, 55, 60, 116, 117, 148, 157
    Pollen, 27, 55
    Products from, 4, 14, 15, 118
    Wax, 55
Beef, 9, 59, 63, 64, 68, 69, 106-108, 113, 115, 122, 124-127
Berry, 16, 20, 22, 55-57. *See also* Blackberry; Blueberry; Bramble; Cranberry; Strawberry.
Birdsfoot Trefoil, 133
Black Locust, 88, 90. *See also* Timber; Woodlot.
Blackberry, 3, 14, 15, 26
Blue Gill, 26, 27. *See also* Fish; Fishing Rights; Pond.
Blueberry, 3, 6, 14, 15, 22, 25-28, 31, 51, 54-56, 60, 124, 147, 148, 151, 154, 157
Boron, 60
Botrytis, 74. *See also* Disease, Crop.
Broccoli, 3, 4, 34, 115
Brochure, 18, 24, 25. *See also* Advertising.
Brussells Sprouts, 3, 4
Buffalo, 64-67

Cabbage, 115, 133
Carrot, 34, 40, 41, 133
Cash Flow, 4, 13, 14, 122.
    Chart, 15
Cattle, 65, 66

Cauliflower, 3, 34
Chain Saw, 33, 87, 91
Chamber of Commerce, 24
Checkout Stand, 17, 19. *See also* Crowd Control; Weigh Station.
Cheese, 117
    Goat, 99, 100
Cherry, 130-133
Chicken, 66, 108. *See also* Poultry.
Chinese Gooseberry, 72. *See also* Kiwifruit.
Chive, 111
Christmas Tree, 117, 154, 157
Cider, 94-96
Citrus, 50, 51, 55, 71
Clientele Membership Club (CMC), 5, 6, 8, 14, 15, 22-30, 33, 34, 57-61, 78, 80, 96, 105-108, 118, 119, 148, 151, 152, 156, 157. *See also* Direct-Marketing; Mail-Order Sales; Pick-Your-Own; Salesmanship.
Clover, 10, 61, 63, 71. *See also* Cover Crop; Green Manure.
    Arrowleaf, 142, 143
    Crimson, 57, 68, 142, 143
    Ladino, 133, 143
    Red, 39, 133
    Strawberry, 143
    Sweet, 89
    White, 131
Codling Moth, 143
Collard, 3, 4, 6, 151
Compost, 111, 140
Computer, 8, 10, 25, 28, 56, 118, 119, 124-126, 151. *See also* Record Keeping.
Container, 7, 16, 17, 26
Cookbook, 30, 119
Cooperative, 2, 3, 9, 99, 100, 124, 154
Coppicing, 88
Corn
    Indian, 154
    Open-Pollinated, 10
    Ornamental, 121
    Sweet, 3, 4, 15, 16, 21, 22, 26, 27, 34, 40, 42, 59, 97, 106, 108, 111, 121, 154
Corporation, 29
Cover Crop, 9, 10, 58, 67, 70, 71, 111, 130, 131, 134, 142, 143. *See also* Clover; Erosion; Grass; Green Manure; Legume; Overseeding; Rye; Vetch.
Cowpea, 58
Cranberry, 117
Crop Mix, 3, 32
Crop Rotation, 9, 10, 41, 111, 133
Cropping System, 10, 11, 16
Crowd Control, 16, 17, 20. *See also* Checkout Stand; Parking; Sign.
Cultimulcher, 40
Cultivation/Cultivator, 9, 10, 33-35, 38-42, 44, 78, 90, 123, 134-136. *See also* Weed Control.

Customer Comfort, 4, 17, 154. *See also* Refreshment; Restroom.

Dandelion, 110
Deer, 60-64, 66, 67, 89, 103, 117, 152. *See also* Hunting Rights.
    Fallow, 60-64, 66, 67, 154
    Meat (venison), 60, 61, 63-65, 118, 157
    Red, 61
    White-Tailed, 61, 64, 66, 117
Direct-Marketing (-Selling), 6, 9, 63, 106-108, 118, 119, 121, 124, 126. *See also* Clientele Membership Club; Mail-Order Sales; Pick-Your-Own; Salesmanship.
Disease Resistance, 10, 117
Disease, Crop, 10, 41, 44, 74, 98, 111
Disease, Livestock, 137-139. *See also* Probiotic.
Disk/Disking, 40, 44, 59, 69, 135, 143
    Disk Hiller, 33
    Harrow, 33, 39
    Offset, 44
    Tandem, 39, 40
Diversification, 4, 9, 10, 22, 56, 121, 122, 124, 150, 151, 154
Domino's Pizza, 153-160. *See also* Monaghan, Tom.
Double-Cropping, 89
Dove, 60-62, 152. *See also* Game Bird; Hunting Rights.
Duck, 64, 154

Eggplant, 41, 43
Egg, 111, 113, 114, 122
Elk, 64-66. *See also* Hunting Rights.
Emu, 64
Equipment, 32-35, 42-47, 69, 89, 96, 104, 123
Erosion, 9, 135. *See also* Cover Crop.
Ethnic Group, 24, 56, 100, 124
Extension Service, 1, 50, 52, 53, 56, 58, 63, 83-85, 88, 96, 98, 144, 149, 150. *See also* Land Grant University; USDA.
External Input, 11
    Chart, 10

Farmers Market, 2, 3, 6, 17, 23, 154
Federal Trade Commission, 121
Fencing, 54-61, 63, 65, 66, 68, 69, 82, 86, 100, 117, 152
Fertilizer, 9-11, 27, 33, 44, 49, 50, 56-58, 60, 71, 72, 78, 151. *See also* Micronutrient; Nitrogen; Phosphorus; Potassium.
Firewood, 85, 86, 88-91, 103
Fish & Wildlife Service, 62
Fish, 4-6, 10, 15, 23, 28, 29, 118, 140, 152, 157. *See also* Bass; Blue Gill; Trout; Fishing Rights.
Fishing Rights, 6, 15, 25, 26, 27, 152, 154. *See also* Pond.

INDEX

Flower, 154
    Cut, 16
    Edible, 79, 80, 109, 110
Foxtail, 134
Fruit Tree, 33, 89, 131
Fungicide, 54

Game Bird, 5, 15, 23, 25, 27, 58, 67, 118. See also Hunting Rights.
Garlic, 3, 15
Geese, 64
Gladiolus, 98
Goat, 68
    Angora, 93
    Dairy, 99, 100
    Meat (chevon), 100
Grape, 3, 15, 26, 27, 32, 50, 51, 147, 154
    Crusher/Stemmer, 32, 33
    Harvester, 32, 33
    Juice, 15, 27, 33, 60, 118
    Muscadine, 54
    Seedless, 124
    Table, 124
Grass, 63-66, 70, 89, 132, 133, 142, 143. See also Cover Crop; Green Manure.
Grazing, 63
    Mob, 59
    Rotational, 57-59, 90, 132
Green Manure, 39, 58, 59, 71, 134. See also Cover Crop; Grass; Legume.
Greenhouse, 80, 82, 97, 98, 110, 124, 134
Green, 3, 4, 15, 27, 34. See also Turnip.
Grow Frame, Solar, 82
Guest Register, 17
Guinea Fowl, 118, 154

Harrow, 134
    Spike-Tooth, 40
    Spring-Tooth, 40
Harvesting, 3, 4, 6-8, 16, 17, 21, 26, 122, 151
Herb, 79, 80, 109, 111, 154
Hickory, 61, 68, 75, 86, 89
Hog, 22, 64, 106-108, 122, 140
Honor System, 22. See also Roadside Stand.
Horse, 5, 9, 23
Hunting Rights, 15, 58, 60-62, 66, 86, 152. See also Deer; Elk; Game Birds.
Hydroponic, 10, 80, 81

Insect Control, 41, 97, 111, 139, 142, 143, 154
    Non-Chemical (chart of 40 common pests), 141
Insurance, 7, 21, 31, 91
Internal Resource, 9, 11
    Chart, 10
Irrigation, 6, 7, 10, 11, 21, 35, 49, 59, 60, 72, 154
    Drip or Trickle, 50, 51, 54, 56, 72, 130
    Portable, 51, 52
    Semi-Permanent, 52
    Solid Set, 51, 52, 58, 59
    Sprinkler, 50, 51

Jerusalem Artichoke, 3, 4

Kale, 4, 6
Kiwifruit, 54, 56, 69-74, 118, 147. See also Chinese Gooseberry.
    Cold-Hardy, 72

Labor, 23, 98, 123, 124
Lacewing, 142
Lady Beetle, 142
Lamb, 3, 4, 22-24, 27, 54, 56-61, 68, 69, 92, 93, 108, 124-127, 148, 154, 157
Lambsquarter, 133
Land Grant University, 1, 2, 7, 25, 52, 59, 148, 149, 151, 154. See also Extension Service; USDA.
Leafhopper, 142
Leafminer, 98
Leafroller, 71
Legume, 10, 11, 57, 58, 70, 71, 73, 88, 89, 131-133, 142, 143. See also Cover Crop; Green Manure.
Lespedeza Sericea, 89
Lettuce, 3, 40, 43, 108, 110
Lime, 57, 58, 60, 78
Llama, 64, 65, 67
Loader, Front-End, 33, 35, 38, 43, 44
Lobok, 16. See also Vegetable, Oriental.
Location, 6, 17, 21, 32, 57, 58, 111, 116, 122
Lupine, White, 134

Mail-Order Sales, 118-121, 125, 126. See also Clientele Membership Club; Direct-Marketing; Salesmanship; United Parcel Service.
Manure, 9, 10, 35, 38, 40, 43, 57
    Pomace, 96
    Poultry, 71
    Rabbit, 56
    Sheep, 56
Maple, 75, 87, 89, 90
    Sugar, 33, 86, 90
    "Super", 90
Marigold, 110
Marketing, 10, 23, 24, 56, 60, 63-66, 69, 83, 84, 104, 106, 108, 116, 117, 123, 124, 153. See also Clientele Membership Club.
Meat, 26, 30, 66, 111, 114, 138
Media Day, 28. See also Advertising.
Micronutrient, 132
Middle Buster, 33
Middlemen, 6, 7, 16, 56, 69, 84, 108, 118, 149, 156. See also Clientele Membership Club.
Millet, 61
Minimum Purchase, 18
Minor Breed, 67
Mint, 111
Monaghan, Tom, 153-157, 160. See also Domino's Pizza.
Mosaic Virus, 98. See also Disease, Crop.
Mowers/Mowing, 69, 117, 130-132, 143
    Flail, 33
    Rotary, 33
Mulberry, Red, 88
Mulch, 9, 41, 90, 97, 98, 111, 130-135, 143
    Machines for laying, 33, 37, 39, 44, 98
Multicropping, 89
Mushroom, 154, 157, 161
    Morel, 89
    Shiitake, 74-77, 80, 154
Muskmelon, 97
Mustard, 3, 81

Nasturtium, 79, 110
Nematode, 73

Newspaper, 18, 20, 22, 24, 102, 107, 116, 124. See also Advertising.
Nitrogen, 9-11, 71, 72, 88, 131, 132, 143. See also Legume.
Nut, 3-5, 15, 23, 26, 27, 78, 89, 118, 120
    Tree, 33, 78, 90
Nutsedge, 134

Oak, 61, 75, 86, 90
    Bur, 88
    Northern Red, 89
    Red, 86
    Swamp, 88
    White, 85, 86, 88-90
Oats, 59, 61, 65, 66, 70, 99, 131, 133, 134, 139
Okra, 15, 16, 18
Orchardgrass, 61, 133
Organic Farming, 8, 69, 71, 72, 81, 106-109, 112-115, 126, 127, 129, 143. See also Regenerative Farming.
Organic Matter, 10, 70, 71, 109
Ornamental, 50, 154
Ostrich, 64, 65
Overseeding, 10, 68. See also Cover Crop.

Packer Wheel, 40, 44
Parking, 4, 17, 18, 20, 21, 109, 123. See also Crowd Control.
Parsley, 110, 111
Peach, 16, 50, 51, 117
Peacock, 64
Peanut, 61, 156
Pear, 87
Pea, 4, 16, 26, 34, 42, 104, 121-123, 134
    Southern, 15
    Sugar Snap, 30
Pecan, 3, 15, 33, 51, 61, 78, 86, 89, 120, 142, 143, 156
Pepper, 16, 41, 43, 78, 97, 124, 153
Persimmon, Oriental, 3, 15
Pest Control, 10, 44, 134, 144, 145
Petting Zoo, 17, 123, 154
Pheasant, 4, 59, 67, 154
Phosphorus, 10, 58, 60, 83, 134
Pick-Your-Own (PYO, U-Pick), 4-8, 16-21, 25, 26, 55, 90, 105, 106, 117, 121-124, 152, 160. See also Clientele Membership Club; Crowd Control; Direct-Marketing; Salesmanship.
Picnic Area, 17, 26, 154
Pigweed, 134
Pine, 61, 86, 88, 89
Pistachio, 3, 78, 120
Planter/Planting, 33-35, 38-41, 44, 69, 83, 131, 136
Play Area, 26
Plow, 9, 33, 35, 44, 143
    Chisel, 39, 135
    Moldboard, 39, 69, 135
Pond, 6, 10, 28, 31, 49, 50, 61, 144, 151, 152, 154. See also Fish; Fishing Rights.
Poplar, 86, 89
Pork, 127
Poster, 18, 24. See also Advertising.
Potassium, 10, 58, 60, 134
Potato, 16, 122
Poultry, 9, 59, 108, 114, 127. See also Chicken; Turkey.
Probiotic, 137-139. See also Disease, Livestock.

Pruning, 74, 88, 89
Publicity, 5, 28, 102, 123, 124. *See also* Advertising.
Pulpwood, 86, 88, 89
Pumpkin, 117, 121, 123, 124, 154

Quail, 4, 28, 59, 118, 154

Rabbit, 3, 5, 14, 15, 23, 24, 27, 28, 56, 118, 124, 152, 154
Raccoon, 64, 86, 89
Radio, 5, 18, 116, 124. *See also* Advertising.
    CB, 19
Ragweed, 68, 133
Raspberry, 3, 14, 15, 23, 26, 111, 116, 117, 121, 122, 124, 154
Recipe, 18, 29, 30, 60, 120
Record Keeping, 8, 28, 56, 124-126. *See also* Computer.
Recreation, 5, 7, 10, 23, 27, 123. *See also* Crowd Control; Picnic Area; Petting Zoo; Rural Atmosphere; Rural Vacation.
Refreshment, 17. *See also* Customer Comfort.
Regenerative Farming, 10, 11, 90. *See also* External Input; Internal Resource; Organic Farming.
Restroom (toilet), 21, 26. *See also* Customer Comfort.
Ridge Tillage, 9, 135, 136
Roadside Stand, 2, 3, 17, 18, 94. *See also* Honor System.
Root Crop, 3, 4, 41
Rotary Hoe, 42-44
Rotary Tiller, 33
Rotovation, 39
Row Cover, 97, 98
Row Spacing, 17, 34, 42, 73
Rural Atmosphere, 17, 23, 101, 124, 154
Rural Vacation, 101-103, 154
Rye, 133, 134
    Annual, 61, 131
    Perennial, 130

Salesmanship, 17. *See also* Clientele Membership Club.
Sawmill, 87-89. *See also* Timber; Woodlot.
Scale, 71. *See also* Disease, Crop.
Seed Crop, 10, 57
Seedbed, 33, 39, 40, 42, 83

Sheller/Shelling, 4, 33, 105, 122, 123
Sign, 31, 116, 123. *See also* Crowd Control.
Slaughterhouse, 57-60, 63, 66, 125, 126
Smother Crop, 10
Soil
    Compaction, 18
    Conservation Service, 62
    Fertility, 10, 52
    Fumigation, 73
    Preparation, 38
    Testing, 52, 53, 57, 58, 60
Soybean, Edible, 104, 105
Spinach, 3, 4, 34, 110, 133
Sprayer, 33, 44, 123
Spreader
    Fertilizer/Lime, 33
    Manure, 33, 43, 44
Squash Blossom, 110
Squash, 16, 97, 98
Squirrel, 60-62, 89, 90, 152
Storage, 16, 18
    Tanks (for juice), 32, 33
Strawberry, 3, 4, 15-22, 26, 50, 54, 74, 117, 121, 122, 124, 151, 154
Succession-Planting, 17
Sumac, 61, 68
Sunflower, 61, 131, 133, 134
Supermarket, 6-8, 25, 26, 69, 95, 112-114, 120, 121, 124, 152
Sweet Potato, 3, 4, 15, 25, 26, 28, 33, 34, 55, 60, 118-120, 151, 154, 156
    Digger, 33
    Washer/Grader, 33
Syrup, Maple, 89, 109, 117

Tarragon, 79, 111
Telephone, 4, 7, 18, 118, 123. *See also* Answering Machine.
Television (TV), 5, 28, 116. *See also* Advertising.
Thinning, 41, 84, 85, 88
Thrips, 98
Tillage, 35, 37, 39, 133-135. *See also* Plow; Ridge Tillage.
Timber (lumber), 83-85, 86-89. *See also* Veneer; Woodlot.
Tomato, 15, 41, 43, 50, 78, 97, 98, 108, 114, 115, 124, 133, 157
Tractor, 33-35, 38, 39, 42, 44, 46-49, 57, 69, 91, 123, 124, 153
    Brand Names/Specs, 36-38

Transplanter/Transplanting, 33, 34, 39, 43, 78, 83, 97, 98, 123
    Finger-Type, 43
    Mulch, 43
Trap
    Insect, 139, 140, 142
    Reptile, 144, 145
Tree Guard, 132
Tree Shaker, 33, 78
Trellising, 54, 70, 73
Trout, 152. *See also* Fish; Fishing Rights.
Turkey, 64. *See also* Poultry.
Turnip, 3, 4, 6
    Green, 151
    TYFON, 3

United Parcel Service (UPS), 120. *See also* Mail-Order Sales.
USDA, 1, 2, 7, 25, 57, 64, 92, 99, 106, 142, 148, 150, 151, 153, 154. *See also* Extension Service; Land Grant University.

Veal, 113
    Free-Range, 112, 114
Veneer, 86, 88, 89. *See also* Timber; Woodlot.
Vetch, 10, 142, 143. *See also* Cover Crop; Legume; Overseeding.
Vine Crop, 43, 51, 57, 69-72, 74, 97, 131

Walnut, 3, 68, 87, 89, 100
    Black, 85, 86, 90
    English, 61, 78, 120
    Tree, 84
Watercress, 81-83
Waterfowl, 10, 67, 144
Watermelon, 97, 139
Weed Badger, 32
Weed Control, 10, 32, 34, 38, 40-42, 57-60, 71, 83, 97, 98, 130-136. *See also* Cultivator.
Weigh Station, 123. *See also* Checkout Stand.
Windbreak, 73
Wine, 33, 156
Woodlot, 10, 60, 61, 83-89, 100. *See also* Timber; Veneer.
Wool, 56, 65, 92-94

Zebra, 64, 66